W9-BLM-077

American Uprising

American Uprising

The Untold Story of America's Largest Slave Revolt

* * *

Daniel Rasmussen

HARPER

An Imprint of HarperCollins*Publishers*

www.harpercollins.com

FIRST EDITION

Maps by Nick Springer, Springer Cartographics LLC.

Library of Congress Cataloging-in-Publication Data
Rasmussen, Daniel.
American uprising : the untold story of America's largest slave revolt / Daniel Rasmussen. — 1st ed.
p. cm.
Includes bibliographical references and index.
ISBN 978-0-06-199521-7
1. Slave insurrections—Louisiana—New Orleans Region—History—19th century. 2. New Orleans Region (La.)—History—19th century. 3. Slavery—Louisiana—New Orleans Region—History—19th century. 4. African Americans—Louisiana—New Orleans Region—History—19th century. 5. New Orleans Region (La.)—Race relations. I. Title.
F379.N557R37 2011
976.'03—dc22 2010017855

---

11 12 13 14 15 OV/RRD 10 9 8 7 6 5 4 3

CONTENTS

American Uprising

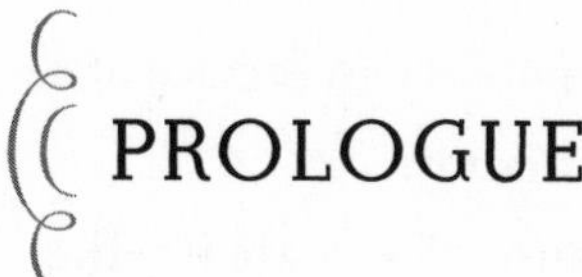

PROLOGUE

THOUGH THE CAUSE OF EVIL PROSPER,
YET 'TIS TRUTH ALONE IS STRONG.

James Russell Lowell

In 1811, a group of between 200 and 500 enslaved men dressed in military uniforms and armed with guns, cane knives, and axes rose up from the slave plantations around New Orleans and set out to conquer the city. They decided that they would die before they would work another day of backbreaking labor in the hot Louisiana sun. Ethnically diverse, politically astute, and highly organized, this slave army challenged not only the economic system of plantation agriculture but also the expansion of American authority in the Southwest. Their January march represented the largest act of armed resistance against slavery in the history of the United States—and one of the defining moments in the history of New Orleans and, indeed, the nation.

In the cane fields outside of the city, federal troops teamed up with French planters to fight these slave-rebels—an unholy alliance between government power and slave-based agriculture that would come to define the young American nation as a slave country in the years leading up to the Civil War. Over the course of the conflict, these powerful white men committed unspeakable acts of brutality in service of a stronger nation and a more vibrant economy.

Because of that brutality, and because of a shared belief in the importance of a specific form of political and economic development, these government officials and slave owners sought to write this massive uprising out of the history books—to dismiss the bold actions of the slave army as irrelevant and trivial. They succeeded. And in doing so, they laid the groundwork for one of the most remarkable moments of historical amnesia in our national memory.

While Nat Turner and John Brown have become household names, Kook and Quamana, Harry Kenner, and Charles Deslondes have barely earned a footnote in American history. Though the 1811 uprising was the largest slave revolt in American history, the longest published scholarly account runs a mere twenty-four pages.

This book redresses that silence and tells the story that the planters could not and would not tell—the story of political activity among the enslaved. What follows is the first definitive account of this central moment in our nation's past—a story more *Braveheart* than *Beloved*: an account of the planning, execution, and suppression of a furious uprising, set in a plantation world far removed from the Virginia of Nat Turner or the sun-drenched plantations of *Gone with the Wind*. This is a story about slave revolutionaries: their lives, their politics, and their fight to the death against the planters and their militia. Above all, this is a story about America: who we are, where we came from, and how our ideals have at times been twisted and cast aside for the sake of greed and power.

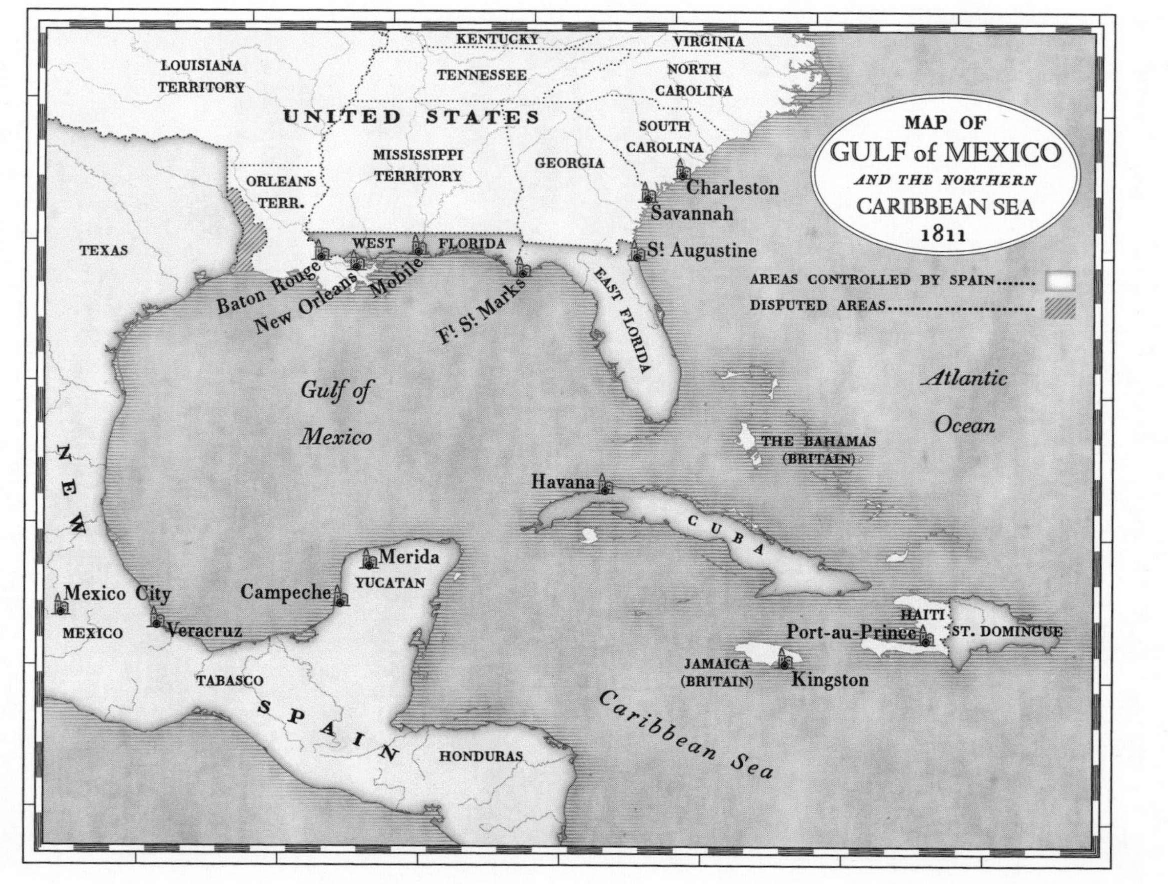
MAP OF
GULF of MEXICO
AND THE NORTHERN
CARIBBEAN SEA
1811
AREAS CONTROLLED BY SPAIN
DISPUTED AREAS
UNITED STATES
KENTUCKY
VIRGINIA
LOUISIANA TERRITORY
TENNESSEE
NORTH CAROLINA
SOUTH CAROLINA
MISSISSIPPI TERRITORY
GEORGIA
ORLEANS TERR.
Charleston
Savannah
St. Augustine
TEXAS
WEST FLORIDA
EAST FLORIDA
Baton Rouge
New Orleans
Mobile
Ft. St. Marks
Gulf of Mexico
Atlantic Ocean
THE BAHAMAS (BRITAIN)
Havana
CUBA
NEW SPAIN
Merida
YUCATAN
Campeche
Mexico City
MEXICO
Veracruz
TABASCO
HONDURAS
HAITI
ST. DOMINGUE
Port-au-Prince
JAMAICA (BRITAIN)
Kingston
Caribbean Sea

January 6, 1811

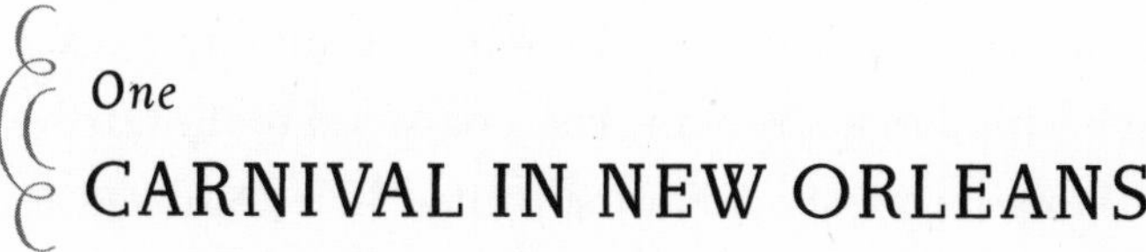

One
CARNIVAL IN NEW ORLEANS

> THE RIVER LEFT GOLD IN THE DELTA. IT WAS GOLD THE COLOR CHOCOLATE, GOLD THAT WAS NOT IN THE EARTH BUT WAS THE EARTH . . . TO TAKE LAND FROM THE RIVER, TO CLEAR IT, DRAIN IT, AND PROTECT IT, REQUIRED AN ENORMOUS OUTLAY OF CAPITAL AND LABOR. FROM THE FIRST THE DELTA DEMANDED ORGANIZATION, CAPITAL, ENTREPRENEURSHIP, AND GAMBLING INSTINCTS. IT WAS A PLACE FOR EMPIRE.
>
> *John M. Barry,* Rising Tide

Down from the mountains of Canada, through dozens of tributaries and smaller rivers, the waters of the American Midwest find their outlet to the sea in the great American Nile: the Mississippi. The river snakes seaward, building tremendous momentum as it hurtles around sharp curves and lashes over rocks, constantly colliding against its own wide banks. For the last 450 miles before the river reaches the ocean, the river bed lies below sea level, and the water has no reason to flow. The water simply tumbles over itself, spitting and gurgling past Natchez and down to New Orleans.

From New Orleans, the river flushes out into the Gulf of Mexico, carrying the continent's commerce into an ocean world rich with ports—from the coast of Africa to the Caribbean to the eastern seaboard of the United States. Situated at the mouth of the river, New Orleans is the prime entrepot of the American Midwest. In the nineteenth century, the city was of central strategic and commercial significance, for through the city, as Thomas Jefferson noted, "the produce of three-eighths of our territory must pass to market." New Orleans was the point at which the commercial farming zones of the Mississippi River valley met the world of Atlantic trade.

Just a few miles north and west of the city, only a few miles up the Mississippi River, began one of the wealthiest and most fertile stretches of agricultural land in North America: Louisiana's famed German Coast. Germans had originally settled the area before being overwhelmed by French immigrants and, in many cases, frenchifying their names and their culture in order to fit in with the new dominant group. Past the gentle slope of the levee stretched green fields as far as the eye could see on both shores. Magnolias, orange trees, and thick oaks sprouted from the sweeping lawns, and Spanish moss hung from the branches. Shielded behind the lush proliferation of gardens stood gorgeous plantation homes built in colonial style, with soaring roofs and columned porticoes.

About twenty miles from New Orleans, smack-dab in the

center of the German Coast, stood the Red Church, a barnlike building with long glass windows, surrounded by a fenced-in cemetery and presbytery. The aristocratic first families of Louisiana streamed out from the front doors.

It was Epiphany Sunday, January 6, 1811, and the planters and their families were buoyant with excitement, undampened by any undue religious solemnity. The local priest was not known for conducting a particularly spiritual Mass. "The social status of the parish priests at the time was not very respectable," wrote one French official. "Adventurers, gluttons, drunkards, often unfrocked monks, they were asked but one thing by their parishioners—that they be, as was said, 'good natured.'"

Good nature was inescapable that Sunday. The previous year's sugar crop was in, and the planters had much to look forward to. Epiphany marked the beginning of a month-long Carnival season. Until its ritual ending at Mardi Gras, all-night parties, mixed-race balls, and constant gambling would occupy the planters' lives. The local French newspaper *L'Ami des Lois* noted plans for an opera and ball at the end of January, and advertised the services of a French-trained dancing master from Haiti and a Parisian hairdresser eager to help the planters and their wives survive the hectic social calendar. Conversations abounded with joy and optimism as the citizens celebrated the unprecedented success of the 1810 harvest.

Several planters, dressed in gloves, hats, and cravats, strolled toward the plantation of Jean Noël Destrehan, a slender Frenchman with dark hair and brown eyes. The roads bustled with carriages and horses; men and women strolled along the levee in their Sunday finery; and slaves hunted and played games in the fields. The planters had a busy day ahead of them. Destrehan, whom one contemporary described as the "most active and intelligent sugar planter in the country," would host lunch at his mansion.

The Destrehan mansion, which survives to this day, was a French Colonial manor, boasting Tuscan pillars tapered into columnettes that upheld a wraparound porch elevated fourteen feet off the ground. The brick-walled first floor was primarily for storage, and the family lived on the second floor. With hardwood floors and twelve-and-a-half-foot-high ceilings, the two- by three-room house was luxurious and comfortable, designed for the enjoyment and display of wealth. Over Madeira and brandy, Destrehan's beautiful female slaves would serve a five-course meal, replacing the tablecloths between each course. Toasts and songs with traditional refrains would punctuate the fine dining as the planters exchanged small talk and gossip.

Amid the oak trees, Spanish moss, and long plantation fields, the planters had developed an unusually tight-knit society. They hosted each other at elaborate dinners, dances, and other entertainment. When planters intermarried, their children started their own plantations. By the early eighteenth century, the Deslondes and Labranche families owned two plantations each, while the Trépagnier and Fortier families owned three plantations each along the German Coast. The plantation homes were symbols of the immense wealth and profit accumulated on the Mississippi Delta. In scale and grandeur, they were unparalleled in the United States.

The meal at the Destrehan plantation was an intimate prelude to the wild and frenzied parties to come that night—

and most nights for the next month. Their repast finished, the planters would take carriages or boats into New Orleans for the evening's celebration at the King's Ball. Slave coach drivers would guide their masters' polished vehicles along the road that ran along the river and the levee, curving gently with each bend in the river, before entering first the Garden District and then the center of the city where the parties would be held.

Every year on January 6, the King's Ball marked the kick-off of the Carnival season that culminated in Mardi Gras. The ball featured the cutting of a cake in which a bean had been hidden. Amid dancing and music, the revelers would celebrate the election of a King and Queen of the Twelfth Cakes, and drink prodigious amounts of alcohol.

After coffee, the dancing would begin. The guests danced boleros, gavottes, English dances, French dances, and gallopades. Elegantly dressed young men smoked and gambled at tables spread around the outside of the room. "You never saw anything more brilliant," wrote a French colonial official fresh from Paris. Slaves brought in supper at three o'clock in the morning, serving gumbos and turtle to the assembled guests at two large tables seating a total of seventy people. After supper, the reveler's took to dancing and gambling again, not leaving until after sunrise. Throughout the month of January and the beginning of February, Destrehan and his fellow planters would devote themselves almost entirely to dancing and gambling—just as their parents and their parents' parents had done.

These bacchanalian traditions dated back to the first French settlements in the area. As early as the 1740s, the Marquis de Vaudreuil had constructed a sort of miniature Versailles in the midst of the earth and log ramparts. In the poor colony on the outskirts of the empire, a place that contained fewer than 800 white males, Vaudreuil took it upon himself to host dances and elaborate banquets, and even to bring in a dancing master from Paris named Bébé (Baby) to teach the next generation to dance.

Jean Noël Destrehan's relatives had been there, as had many of his friends' ancestors. Dressed in wig, satins, and lace, Destrehan's grandfather, Jean Baptiste Honoré Destrehan de Beaupré, had arrived in New Orleans in 1721, bringing the royal Destrehan bloodline to the shores of the Mississippi. He had served as the first treasurer of the new colony. And despite living in a colony far removed from Paris, the Destrehans maintained their elite traditions and family reputation.

Jean Noël Destrehan himself had traveled to France for his education before returning to the New World to marry and run the family plantation. In a few short years, Destrehan impregnated his wife fourteen times, producing eleven children who survived infancy. In the early 1800s, he had to add two additional wings to his mansion to accommodate his hearty brood—a construction project amply paid for with money earned from his profitable career as a sugar planter.

Universally regarded by his fellow planters as a wise and generous man, Destrehan became the de facto spokesman for the motley set of French expatriates. Cultivated and elegant, he was a symbol of the planters' firm belief in the power and supremacy of French civilization. These settlers had chosen to abandon the luxuries of Enlightenment-era France for the wild tropics of Louisiana because they believed that here they could make vast fortunes and become men of wealth and status. Destrehan—and his lavish lifestyle of banquets and parties—provided living proof that this dream could become reality. Like many of the planters, Destrehan was rapidly becoming very rich from growing sugar.

As a business proposition, sugar planting was relatively simple: maximize quality and quantity of sugar cane output through the use of slave labor to exploit the natural landscape. The primary investments of sugar masters—land and slaves—achieved higher rates of return in New Orleans than anywhere else in the United States. "Those who have attempted the cultivation of the Sugar Cane are making immense fortunes with the same number of hands which in Maryland and Virginia scarced suffice to pay their annual expences," wrote a correspondent for the Louisiana *Gazette*.

But making that business proposition a reality came at immense human cost. Force, or the threat of force, was as necessary an investment as land in making a successful sugar plantation. For slaves would not work without coercion. The planters seemed to focus their attention less on the methods and tremendous injustices of their chosen lifestyle and more on the results; perhaps this was the only way to rationalize the tremendous risks. Yet this heavy investment in violence created a fundamental risk: that the violence would backfire, wreaking uncontrollable havoc on the architects of this brutal system.

Habituated by time and custom to these rigid power relationships, the planters saw no contradiction between their lifestyles and the system of enslavement. By taking credit for the work of people they considered to be their property, they told stories about their accomplishments and their plantations without reference to those who made it all possible. The planters discussed and showed off their beautiful mansions, their lives of leisure, their abundance of slaves, their well-constructed buildings. They built reputations as manly independent patriarchs, as gentlemen farmers, and as powerful aristocrats. To these men, slavery signaled status and wealth, not immorality or danger.

Destrehan and his friends considered their wealth the fruit of their own labors. In their minds, they were the ones who had worked hard, and they were the ones who should reap the rewards. Many criticized them for their lavish lifestyles, but the French planters had little patience for such attacks. "We could not imagine what had produced the idea of our effeminacy and profusion; and the laborious planter, at his frugal meal, heard with a smile of bitterness and complaint the descriptions published at Washington of his opulence and luxury," Destrehan wrote.

And he did work hard. During most of the year—with the exception of the Carnival season in January when there was less to do on the plantation—Jean Noël Destrehan maintained the strict daily routine of a typical French sugar planter. Awakening at sunrise, he made his appearance on the piazza of his house and took his coffee, toast, and tobacco pipe. Then he met with his slave drivers to plan the day's work, approve specific punishments, and hear about the state of the plantation. As the bells rang to signal the beginning of work, Destrehan strutted out into the fields in his morning dress (holland trousers, white silk stockings, red or yellow

Moroccan slippers, and a cotton nightcap to keep off the hot Louisiana sun). His slave mistress might accompany him on this walk, offering him a morning Madeira and pipe to refresh him. Destrehan, wrote one French official, "was there all the time, following all of his operations. Woe to anyone who would disturb one of his Negroes at this time, or his horses or oxen! Obliging though he usually was, this would have been like stabbing him in the back."

Destrehan believed that there could be no German Coast without slavery. He believed that without chattel slavery, "cultivation must cease, the improvements of a century be destroyed, and the great river resume its empire over our ruined fields and demolished habitations." Indeed, slaves were an absolute necessity—the very foundation of this strange frontier world. By 1810, slaves constituted more than 75 percent of the total population, and close to 90 percent of households owned slaves.

In fact, the planters used the very strangeness of the land—with its heat and disease and wild, uncontrollable river—to justify the mass importation and forced labor of African slaves. Destrehan saw Africans as uniquely matched to the hot weather and tough work. "To the necessity of employing African laborers, which arises from the climate and the species of cultivation pursued in warm latitudes, is added a reason in this country peculiar to itself," he wrote. "The banks raised to restrain the waters of the Mississippi can only be kept in repair by those whose natural constitution and habits of labor enable them to resist the combined effects of deleterious moisture and a degree of heat intolerable to whites." Slaves were the planters' defense against the great river, their weapons in a contest between civilized man and untamed nature.

Destrehan did not mention the spiked iron collars, cowhide whips, and face masks that he and the other planters used to encourage the slaves' "natural habits." Though the planters had no difficulty reconciling the wealth they enjoyed and the price the slaves paid, the region's black laborers did. By aborting their own children, poisoning livestock, lighting fires, and escaping to the cypress swamps, the slaves struggled to dilute, deflect, and if possible demolish slaveholders' authority. Even open revolt was not beyond question. While it was a card that slaves played only rarely—planters tended to take a dim and deadly view of armed rebellion—the German Coast teemed with violent possibilities. The planters' world rested on a powder keg ready to be ignited by the smallest of sparks. Unbeknownst to those who crowded the ballrooms and attended the season's festivities, that spark had already been lit.

Two

PATHS TO SLAVERY

African and African-descended slaves, Native Americans, free colored people, white women in flaring yellow-and-scarlet gowns, men with caps and hats, plump Americans from the East Coast, and skinny Spanish settlers jostled between the twin rows of the market that extended along the New Orleans levee. While a few ambitious salesmen had erected tables with canvas awnings, most clustered on the ground laying their wares on pieces of canvas or palmetto leaves. Everything one could imagine was on sale—from wild ducks to bananas, oranges and sugar cane to trinkets, tinware, dry goods, and books. The hollers of the salesmen, the shouts of the sailors, and the constant chatter of barter and exchange filled the air with noise—there could not have been fewer than 500 people crowded along the riverbank. And as on every Sunday, there were more black faces than white. Under French custom, many slaves in Louisiana were allowed to raise vegetables and poultry, the surplus of which many exchanged in the region's markets for clothes, money, and basic necessities of life. They also came to mingle, socialize, and dance.

The Place d'Armes, the wide central square of New Orleans, opened past the markets. To the left, the towers of the Cathedral of St. Louis surged skyward; to the right, a hotel with narrow galleries along the first and second stories shaded the passersby.

As the planters celebrated Epiphany, five or six hundred African men and women from New Orleans and its environs gathered in the Common. Some of their faces bore the ritual facial scars and filed teeth of the Kongo, others the tattoos of the Asante kingdom, while others born in Louisiana or the Caribbean simply bore the scars and calluses of a life spent harvesting sugar. Here in a public square not far from the center of the city, these men and women were able to temporarily forget the harsh conditions of their Louisiana lives.

The participants formed dancing circles around one central ring, where two women danced languorously, holding handkerchiefs with the tips of their fingers. An old man sat on top of a cylindrical drum about a foot in diameter, beating the canvas quickly with the edge of his hand. Several others held drums between their knees, producing incredible explosions of percussive sound. A grizzled man of not less than eighty years of age played an African stringed instrument that extended from the ground to over his head. Women sang African folk songs while walking rhythmically around the drummers. The sounds of Africa rocked the Crescent City.

The square burst with color. The men came wrapped in traditional coastal African garb, brightly dyed robes wrapped around their otherwise naked bodies. The women boasted the latest fashions, sporting clothes made of silk, gauze, muslin, and percale. The flags of different African tribes, regions, and ethnicities flew above the celebrants.

On this day, the day of the anointment of the planters' Carnival king, this festival would also celebrate the crowning of a new leader. Many times these leaders had been chiefs or kings in Africa. The crowning of the leader was the climax of the festival. The crown looked like a series of brightly colored paper boxes, tapering upward like a pyramid, with two tassels hanging downward from the pinnacle. Ensconced on his throne and crowned with this strange headgear, the king would take ceremonial command. "He wags his head and makes grimaces," one visitor to New Orleans wrote. "He produces an irresistible effect upon the multitude."

These dances were probably the largest African gatherings for hundreds of miles; because of its French and Spanish past, New Orleans still allowed large groups of black people to meet—unlike most of the rest of the United States, where such gatherings were banned for fear of the discontent that might surface. These celebrations, religious and profane, were a long tradition in New Orleans, dating from at least the 1770s. These meetings served as a means of exchange, both cultural and economic. But they also served another role, as a breeding ground for slave conspiracies. And that Sunday, the circuits of secret slave communication buzzed with signals: a plot was afoot.

* * *

Two men, Kook and Quamana, were in large part responsible for activating these African channels with revolutionary activity. Like the participants in the dance that January Sunday, Kook and Quamana were Africans. Their names suggest that these two men were Akan, children of a warlike African empire in the height of its glory. Their names were anglicizations of the names Kwaku and Kwamina in the Akan dialects of Twi and Fanti. The name Kook was the name associated with the spider—the classic trickster Kwaku Anansi of Akan folklore—and meant that Kook was born on a Wednesday. The name Quamana was assigned to men born on Tuesdays and was associated with the ocean.

They brought with them from Africa the memories and stories of the powerful and warlike empire in which they most likely grew up. The Asante kingdom controlled large swaths of land around the Bight of Benin stretching across present-day Nigeria, Benin, Togo, and Ghana, a distance of about 300 miles. The Asante kingdom was a military union made up of a diverse range of local tribes. Many Akan slaves brought to America had been *okofokums,* common soldiers trained to fight in the massive armies of the West African continent. If Kook and Quamana were indeed Akan, they would have at the very least known how to use a weapon and how to fight—these were things Akans learned from birth.

According to their master's estimates, Kook and Quamana were born around 1790—a time when war was ravaging the African continent. In the Lower Guinea region, the Oyo empire, the kingdom of Dahomey (known for having palaces decorated with human skulls), and the smaller states of the coast fought for regional supremacy, enslaving and selling prisoners of war to European traders at coastal forts. In the kingdom of the Kongo, a series of civil wars that had begun in 1665 mobilized large numbers of troops, with guns and horses, who fought across the African plains. The violence peaked around 1781, when 30,000 Kongolese warriors stormed the fort at São Salvador under the command of King José I. In the Gold Coast, the Akan were driving toward the coast with massive armies, threatening the European vassal states that supported the Atlantic slave trade. Kook and Quamana were born, grew up, and were sold into slavery amid these international conflicts, children of war fueled by a fast-growing Atlantic economy.

Kook and Quamana arrived in port in 1806 as part of a vast wave of forced migration. That year, ten separate slave ships arrived in the port of New Orleans, bringing a total of between 1,000 and 1,500 slaves to the slave marts of Louisiana. Most of these slaves arrived on board the brig *Carolina* or the schooner *United States,* both of which made multiple trips to and from Charleston, South Carolina. From about 1770 to the legal closing of the slave trade in 1808, slave traders brought an estimated 24,000 to 29,000 slaves to New Orleans, with roughly half of those slaves arriving in Louisiana after the United States took control in 1803. These immigrants were almost all Africans during the Spanish period, and three-quarters African during the American period. During the American period, about 40 percent of slaves came from West Central Africa, 20 percent from Senegambia, and the remainder from the Bight of Biafra, the Gold Coast, the Bight of Benin, and the Windward Coast. The Asante people primarily came from the Bight of Benin.

On its way to New Orleans, Kook and Quamana's slave ship most likely stopped in Charleston, South Carolina. Like Jamaica and Havana, Charleston functioned as a way station in the slave trade. Some slaves might have stayed for a year or two in these entrepots to be broken in; others might have stayed for a few weeks in a slave market, while still others might never even have gotten off the boat and just waited as their ship resupplied. The Christian slave traders and planters baptized roughly a third to a half of the slaves who arrived in New Orleans during the period in which Kook and Quamana arrived, implying that roughly this portion came direct from Africa, while the rest had been seasoned and baptized elsewhere.

To the ship captains, the bodies packed in the hold were just another type of cargo, but down in the deep darkness below the decks were people with powerful memories of their homeland, and more often than many planters might hope, experience in warfare. To survive this journey, from imprisonment and sale in Africa to baptism in the great Cathedral of New Orleans, required a strong constitution and an ability to witness and endure ferocious and traumatic violence. Death was endemic to the system of enslavement. Forty percent of those captured died even before boarding a slave ship. Another 10 percent died either on the Middle Passage or shortly after arrival in the New World. Only 30 percent of slaves captured in Africa would survive past the third or fourth year of labor.

Kook and Quamana were among the survivors. But beyond these details and statistics, what were their lives really like? What impact did that experience of exodus and diaspora have in shaping these men into the people they became? The stories of most who made these journeys will forever go untold; some few broke the silence, and their voices—and their histories—provide a window into Kook's and Quamana's experiences.

Perhaps the best lens into the routes to slavery is a book published by a man named Olaudah Equiano in 1789. Equiano claimed to have been born in Nigeria and brought over from Africa in a slave ship, though records suggest he was actually born in South Carolina. While Equiano's story may not have been his own, the story nevertheless offers us a glimpse at how a slave experienced, imagined, or heard about the Middle Passage and the transition from Africa to America—a journey better described by a former slave than any white slave trader or abolitionist.

* * *

By his telling, Equiano was born deep in the interior of Africa in the fertile Essaka valley. Growing up in the wooded plains, Equiano had never heard of white men, or Europeans, or of the sea; he knew of the king of Benin, but even that authority was more legendary than concrete. The chiefs and elders of the village were the true authorities. These elders bore the embrenche, a ritual scar made by cutting the skin across at the top of the forehead down to the eyebrows, creating a ridged welt that signified high distinction. Equiano too would receive such a mark when he came of age. Eating goats and poultry, plantains, yams, beans, and corn, the villagers resided in thatch-roofed family dwellings separated by moats or fences. Each building was covered on the inside with a composition of cow dung to keep out the tropical insects. Equiano and his family slept on elevated beds covered with animal skins.

"We are almost a nation of dancers, musicians, and poets," Equiano recalled. Every marriage, birth, victory in war, and other public event would be celebrated with public dances. As betrothed virgins played instruments resembling the guitar and the xylophone, four divisions would dance apart or in succession: the married men came first with ritual warfare, while the married women, young men, and maidens followed with scenes of real life or myth. Men and women both wore long pieces of dyed-blue calico wrapped loosely around the body, while the women of distinction wore golden jewelry as well.

But all was not calm and peaceful in Essaka, as constant wars rippled through the territory. Every man, woman, and child quickly learned to use a variety of weapons. The village was stockpiled with guns, bows and arrows, swords, javelins, and tall shields. "Our whole district is a kind of militia," Equiano wrote. Entering war, the militia would carry a red flag or banner. The winners would enslave or kill the losers. Often the women were kept as slaves to the Africans, while the more troublesome men were sold away in lieu of being put to death. War was so constant that Equiano remembers every man and woman going out into the fields armed lest the enemy surprise them amid their plantings and harvestings.

When the adults went out into the fields to farm, the children banded together to play games. The village of Essaka was not a safe place for children, however, and one child would be appointed to watch from the treetops for assailants, kidnappers, or enemy tribesmen who might take advantage of the adults' absence to carry off as many children as they could seize. Some of Equiano's earliest childhood memories were of kidnappers sneaking into the village to bind and make off with his playmates; it would not be long until this was his fate, too.

One day, while the adults were out in the fields, two men and a woman scaled the walls of the village and captured Equiano and his younger sister. Gagging the two children, the kidnappers rushed them as quickly as possible out of the village. Overpowered, Equiano had no chance to escape. The raiders marched him and his sister far into the woods, stopping only briefly to sleep in a small cabin in the woods. After another day's travel, they came out onto a road, and Equiano began to scream in the hopes of being rescued. But his kidnappers were rough men and quickly gagged him before tossing him into a sack. He and his sister refused food: "the only comfort we had was in being in one another's arms all that night, and bathing each other with our tears." That comfort too would be denied him, as the kidnappers sold off his sister the next day.

After a long journey, Equiano came to the river; he had never before seen water greater than a pond or spring. "My surprise was mingled with no small fear when I was put into one of these canoes, and we began to paddle and move along the river," he wrote. Within six months of leaving Essaka, Equiano arrived at the seacoast—and saw the looming masts of the slave ship towering over the trees. He was terrified.

"I was now persuaded that I had gotten into a world of bad

spirits," he remembered. With long hair, strange language, and bright complexions, the white men seemed like hellish demons. The other black people about wore dejected looks and grim expressions. They talked anxiously of the fears of being eaten by the white men with their "horrible looks, red faces, and loose hair." Some slaves choked themselves to death by swallowing their tongues, others cut off their fingers to make themselves damaged goods, while still others just gave up the will to live and died. Equiano was horrified and despaired of ever returning to his native land. He watched as those who resisted the white men were savagely beaten, cut, and chained—even white sailors received this treatment on occasion. His initiation into the world of Atlantic slavery had begun.

The slave traders brought the slaves from the beach to the slave ships in canoes that tossed and turned in the waves, upsetting the stomachs of African men and women little accustomed to the sea. The traders bound the slaves' ankles together with iron rods and loaded them two by two into the cramped semidarkness of the slave ship's hold. But the newly enslaved did not give in easily to these new oppressions. As countless captains' logs, sailors' stories, and voyage records suggest, slave ships were notoriously explosive places.

The captains of the slave ship *Diligent* left detailed records that provide a window into life aboard a slaver—and the constant warfare between the traders and the slaves. The first moments with a new cargo of slaves were an incredibly tense time for the crew, who knew that slave revolts were most likely to occur in sight of land. The captain posted a rotating set of sentinels to guard the hatchway to the slave quarters twenty-four hours a day. The crew kept arms stocked on deck and kept the swivel guns on the quarterdeck trained at the main deck near the hatchway. Mealtimes, when the slaves were brought on deck, were the moments of highest tension—the entire crew would turn out with guns loaded to help keep order, preventing both revolt and suicide. Tensions ran high.

The crew saw rebellion in every dark face, fearing constantly lest their ship erupt into violence. Traders knew that children could carry messages and use their sharp eyes to discover loose nails and other potential weapons, that women could use their greater freedom to spy and survey the opportunities for revolt. In fact, studies suggest that slave rebellions were more likely when there were more women on board. The slaves did more than just plot revolt; they also shared valuable information, established networks of communication, and laid the groundwork for the pan-African slave culture of the New World. The informal networks that would sustain these men and women in captivity began to form in the dark holds of the ships. The slaves shared their histories, established bonds of trust and affection, and speculated constantly on what might await them at their destination.

Slave traders made examples of those who did choose to attempt escape or revolt. When one slave on the *Diligent* bit a fellow slave in order to try to spark a revolt, the captain brought the entire cargo of slaves onto the deck. He ordered sailors to tie a rope around the man's chest and hoist him up high into the air. A firing squad then unleashed a volley of shots as the man screamed with anger and agony. His blood spattered onto the faces of the slave onlookers. The captain wrote in his journal that he did this "to teach a lesson to all the others." As the Africans transitioned into their new roles as slaves, they would witness many such lessons.

Once the slave ship reached open ocean, the traders worked to try to keep the slaves in a semblance of health, lest they lose too much of their cargo. The profit was in delivering laborers—not half-dead corpses—to New World markets. Killing and maiming the cargo was really the last thing wanted by ship captains and their sponsors; that they did it anyway demonstrates the violence and fear prevalent on these journeys. To keep the slaves' muscles in shape, the sailors would bring the slaves onto the deck and have them dance. As one sailor played the accordion or guitar, other crew members prodded the shackled men to jump up and down in unison on the pitching deck of the vessel. "It was usual to make them dance in order that they might exercise their limbs and preserve their health," explained one British surgeon on a slave ship in 1789. "This was done by means of a Cat of Nine Tails with which they were driven about one among the other, one of their country drums beating at the same time. On these occasions they were compelled to sing, the Cat being brandished at them for that purpose ... The men could only jump up and rattle their chains, but the women ... were driven among one another." They had other plans for the women. The sailors would strip them completely naked and have them dance unshackled for the amusement of the crew. At nights, the officers went down to the quarters and raped women at will.

After a journey of a few months, depending on weather conditions, the slave ships arrived at the various entry ports of the New World—Charleston, Havana, or Kingston. Prior to their arrival in these slave marts, experts among the crew rubbed the slaves' skins with oil to make them shine, gave them rum to clear their eyes and brighten their countenances, covered their sores with iron rust and gunpowder, and closed their lesions. They wanted the slaves to look as healthy as possible before sale.

Traders would buy and sell the slaves here. Some would be bought to serve in the West Indies, others for transshipment to North America, while still others would serve for a bit in the West Indies before being sent on to North America, having either been bought for the purpose of being broken in and then resold, or simply rejected by the West Indian planters.

Several thousand of the many slaves caught up in the New World ended up in New Orleans. Here, emerging onto the deck, the slaves would have seen a bustling waterfront—ships from all over the world, hundreds of flatboats packed densely together, stevedores loading and unloading goods, sailors shouting instructions—a cacophony of exchange conducted in half a dozen or more languages. The sailors brought the slaves on deck, then transported them in flatboats to the shore. The slave merchants took the terrified Africans from there, marching them in chains through the central square of the city, past the cathedral and the Cabildo, past the sailors' district with its shacks, brothels, and bars, to a huge slave market advertised by hanging signs with names such as "Kenner and Henderson." The merchants packed the slaves into pens the size of home lots surrounded by fences fifteen to twenty feet tall.

* * *

Sometime between May and September of 1806, the brash American James Brown drove his carriage in from his new plantation on the German Coast and parked it outside one of these slave markets, perhaps outside the full-service slave firm run by his fellow Americans William Kenner and Stephen Henderson. Kenner and Henderson had arrived flush with cash from White Sulphur Springs, Virginia, to set up a sugar plantation and merchant firm. The two operated a full-service business that shipped plantation produce to market, provided financing and insurance, bought and sold slaves, and procured building materials and other necessities for planters.

Brown had arrived in Louisiana only a year before, buying plantation land on the German Coast just west of the noble French family the Trépagniers. He had watched the value of his plantation more than double over the course of the year, from $16,000 to over $40,000, as the price of sugar rose and his slaves converted the land for sugar production. Brown was a memorable man. A contemporary from Kentucky described him as a "towering & majestic person, very proud, austere & haughty in fact repulsive in manner, and . . . exceedingly unpopular."

In the hot weather of 1806, Brown was not in New Orleans to win a popularity contest. He was there to make money. Brimming with ambition, he came to the slave market that day to buy slaves who would make him rich.

Here, for the first time, Brown set eyes on Kook and Quamana, the first a mere fifteen years old, the latter twenty-one. With finished floors and beautifully painted walls, the showroom could hold a hundred slaves. Kook and Quamana would have watched anxiously as Brown strolled through the aisles of the mart inspecting each individual slave. The slaves tried to imagine the characters of their potential owners—to gauge what their fates might be at the hands of these men. The planters dressed up to buy slaves, in full black suits or multicolored pants, stiff top hats or wide-brimmed shapes, some with ties or jewelry, others with canes or walking sticks. If James Brown was as repulsive as his white contemporaries described him, he must have cast a truly terrifying figure to the two African men quivering in the corner. That day, James Brown decided to purchase both Kook and Quamana. He paid $700 for Kook and $600 for Quamana (about $11,000 each in modern values).

But Kook and Quamana were not fated to become stock characters in Brown's plantation drama. Soon after their arrival in New Orleans, they chose to reject their new status as slaves and to begin plotting a ferocious rebellion—a rebellion that they hoped would bring them back into New Orleans not in chains but in triumph. Amid the swirling diversity of New World slavery, Kook and Quamana slowly began to identify and cultivate a network of like-minded slaves, a network they would have had to hone through day-to-day interactions, without attracting the notice of the keenly observant planters.

Kook and Quamana must have taken advantage of discreet meetings in cabarets in the city, in the homes of free blacks, and in the slave quarters. Even the planters were aware of the extent of these gatherings—though they supposed they were recreational, not revolutionary. On the German Coast, the home of Joseph the Spaniard was a known location for slaves to drink and congregate on the weekends. In 1763, the Spanish attorney general had complained about illicit tavern keepers like Joseph: "While furnishing drink they incite them to pilfer and steal from the houses of their masters," he wrote. "[The slave] would not be violent if he did not find in these secret taverns the means to satisfy his brutal passions; what hidden pernicious disorders have resulted." The Spanish, with their long experience, knew how dangerous these uncontrolled slave activities could be.

Yet these officials did not ban the dances—nor could they restrict the constant movements of slaves between plantations and around New Orleans. Slaves served as messengers and deliverymen, and they were responsible for relaying goods and news from plantation to plantation at their masters' behest. They traveled into New Orleans to their masters' town houses, and they traveled to the marketplace to sell goods. Slaves were also allowed to travel for family reasons. Many male slaves had wives at other plantations, whom they were allowed to visit on the weekends. It was unusual for a slave to spend his or her entire life on one plantation. The masters frequently rented out their slaves to other planters for a fixed sum of money. Whenever a planter died or a son became old enough to start a plantation, slaves would be redistributed, moving from place to place around the German Coast. And as they moved, they built contacts and relationships—a network of acquaintances and trusted friends they could use to spread gossip, news, political ideologies, and, in the months leading up to January 1811, a plan for revolt.

And there was nowhere better to build a revolutionary organization than the dances in New Orleans. Here slaves like Kook and Quamana could talk away from the watchful eyes and listening ears of the white planter class. Though the Americans and French showed no understanding of the possibilities of these dances, the Spanish were well aware of this danger. "Nothing is more dreaded than to see the negroes assemble together on Sundays," wrote a Spanish historian in 1774. "In these likewise they plot their rebellions." In some African cultures, dancing was about more than celebrating. Dancing could double as military training, developing individuals into fit and cohesive groups. In fact, this form of military training was common enough that the Kongolese used the phrase "dancing a war dance" as a synonym for declaring war.

Slaves in the New World were quick to take advantage of festivals and celebrations, using their masters' carelessness and inebriation to plot and often carry out rebellions. In 1812, for example, a group of Akan slaves organized a rebellion in nearby Cuba. The slaves organized the rebellion between Christmas and the Day of the Kings on January 6. The leaders met at taverns, festivals, and other small gatherings, using travel passes and visitation rights to move without being noticed by their masters. At one meeting, the slave José proclaimed that "if they were to be captured, it would not be alive, but dead."

But while the festivals provided the cover for the final meetings, revolts did not crop up overnight. Rather, organizing a successful revolt in the face of tremendous odds and suspicious planters required secrecy, organizational skill, persistence, and above all, trust. No record survives of just what Kook and Quamana said or how they plotted their uprising, but another revolt led by Akan people in New York in 1741 gives us a picture of how they might have operated; as in New York, the 1811 uprising involved a wide diversity of African peoples, drawn from all over the Atlantic world, and of many different languages and nationalities.

An inner circle of "headmen" was responsible for organizing specific communities into insurrectionary cells—for "recruitment, discipline, and solidarity." With the rewards for betraying a revolt extremely high, headmen like Kook and Quamana had to be extremely careful about whom they spoke with; they had to be sure they could be trusted. In New York, headmen had focused on organizing within specific national groups. Slaves were "not to open the conspiracy to any but those that were of their own country," wrote a participant in the New York revolt, since "they are brought from different parts of Africa and might be supposed best to know the temper and disposition of each other." They addressed each other as "countrymen" and used a coded language to feel out other slaves' beliefs and politics. New recruits swore a war oath when they joined an insurrectionary cell. These military oaths were widespread across West Africa and invoked the "primal powers of thunder and lightening" to ensure utmost secrecy and violent camaraderie.

Fortunately for Kook and Quamana, there had been a significant change in New World slavery since the New York uprising. Before 1800, no slave revolt had ever been successful. But in the first years of the new century, a group of slaves on a French island in the Caribbean launched a massive revolution meant to overturn European power and establish a black republic in the heart of the Atlantic. The stories of this daring gambit were well known to Louisiana slaves. The links to revolutionary Haiti were far closer than the planters would have liked. And there is little doubt that Kook and Quamana used the stories of this revolution to inspire and cajole their fellow slaves into joining their planned insurrection.

Three

A REVOLUTIONARY FORGE

Twelve hundred nautical miles east-southeast from New Orleans, the verdant 6,000-foot peaks of the island of Saint Domingue (now Haiti) jut out from the ocean. Freshwater streams flow down from the mountain peaks, forming rivers that fertilize the valleys and plains that lie beneath. The weather of the tropics adds lushness and fertility to the air—greenery bursts out of every crevasse, from under every banana tree, beneath the cloudless blue skies. Mango, orange, and coffee trees grow naturally here. The year was 1791, and the farmland on this tropical island was perhaps the single most valuable property on earth.

Though Haiti is now a very poor country, then the soil of the island yielded untold riches. In 1767 alone, the French colony of Saint Domingue exported 123 million pounds of sugar, two million pounds of cotton, a million pounds of indigo, and vast quantities of hides, molasses, cocoa, and rum. And that was only the start of the island's agricultural boom. Not even 11,000 square miles, the tiny island was a hub of an Atlantic commerce—the jewel in the crown of the French empire.

The island accounted for over 60 percent of France's export trade, and more ships docked in its ports than in Marseilles or any other smaller French port. As French ships brought Caribbean sugar and other cash crops to Europe, they brought back with them the processed goods of Europe—salted cod and other meats, brandy and wine, flour and all manner of refined goods. This valuable colony drove a period of rapid economic growth in France and in Europe more broadly, a period of economic growth that laid the groundwork for the Industrial Revolution. Whole factories—whole towns—grew up to serve this vast New World trade. And as the tall-masted ships sailed back and forth across the Atlantic, Europeans watched their coffers fill with gold and the benefits of trade lift large swathes of the population from peasantry into the middle class.

Yet beneath this story of wealth and riches, behind this tale of progress, lay darker realities. Sugar, cotton, and coffee don't grow themselves. They demand backbreaking, intolerable labor—labor to which no free man would choose to submit. The task of raising cane in the fields of Saint Domingue and harvesting the lush crops of the island fell on the backs of the Enlightenment's greatest and most productive laboring class—African slaves.

After decimating the native population, Europeans imported around half a million slaves from the coasts of Africa to this tiny island over the course of just a few decades. In the process of ripping these men and women from their native homes and transporting them by force to a New World where most died within seven years, this Atlantic trade fueled wars across the African continent, cost untold millions of lives, and, of course, brought unprecedented prosperity to the slave traders and the planters and merchants who depended on them.

That raising crops could be so profitable seems very foreign to the modern eye, but in that day and age the production of sugar was the most profitable form of agriculture. Consider the famed Gallifet plantation, where 808 slaves worked to harvest the New World's most lucrative crop. The master of the plantation once asked, "How can we make a lot of sugar when we work only sixteen hours [per day]?" The answer, he concluded, was "by consuming men and animals." And indeed, the Gallifet plantation did consume men, quite quickly and efficiently. These colonial plantations were as close to a death camp as one could come in the late eighteenth century. Overseers carried swords and whips to punish recalcitrant slaves. Few slaves lived past forty and most died within a few years of starting plantation work.

But as these complex economic relationships played out on the Atlantic, creating a vast network of death and profit, other forces too were at work—forces not amenable to empire or capitalism. Try as they might, the slave owners could not turn people into machines—and people do not submit easily to cruelty and exploitation. One liberal traveler on the island noted the judgment and resentment that the slaves expressed when by themselves. "One has to hear with what warmth and what volubility, and at the same time with what precision of ideas and accuracy of judgment, this creature, heavy and taciturn all day, now squatting before his fire, tells stories, talks, gesticulates, argues, passes opinions, approves or condemns both his master and everyone who surrounds him," the traveler wrote.

Had this traveler been an African, he might have discovered much more. He might have known the meaning of the African chant "*Eh! Eh! Bomba! Heu! Heu! Canga bafio te! Canga, moune de la! Canga do ki la! Canga, li!*" or, in English, "We swear to destroy the whites and all that they possess; let us die rather than fail to keep this vow." That same traveler, had he been an African, might also have been invited to join certain congresses held late at nights in the woods away from the plantations. For in August of 1791, the slaves were plotting to make their chant a reality. In furtive conversations held far from the planters' watchful eyes, the slaves decided that this would be their last summer on French-owned plantations. They would start "a war to the death against the whites." Given all they had suffered, perhaps it was only time.

On the night of August 21, a band of slaves rose up in arms. The first victim was a refiner's apprentice. They caught him in the sugar factory and cut him into pieces with cutlasses. When his screams awoke the overseer, the slave-rebels shot the overseer dead too, before proceeding to the apartment of the refiner, whom they killed in his bed. From there, they traveled from plantation to plantation, raising a force of nearly 2,000 slaves, setting fire to the cane fields, killing white women and men, and burning houses. The fires were visible for miles and miles. Their attacks, reported one planter, "spread like a torrent."

The group of slaves who began this revolt must have known the punishment for a suspected rebel: ritual torture and death, combined with dismemberment to ensure that their souls could not pass into the afterworld. But perhaps they also knew that staying and working in the fields would lead to death just the same—though in a few years rather than a few days, and by exhaustion and malnutrition, not violence. But those who made Saint Domingue's sugar were strong and had inspiring leaders.

The most visible organizer was a coach driver and former slave driver named Boukman, a man known as a religious leader. In the first days of the revolt, Boukman gathered a band of slaves in the woods at a place called Bois-Caiman, where he led the slaves in a religious ceremony. A woman—variously described as having "strange eyes and bristling hair" or having green eyes and being of mixed race—presided with him. "The god of the white man calls him to commit crimes; our god asks only good works of us. But this god who is so good orders revenge," declared Boukman. "He will direct our hands; he will aid us. Throw away the image of the god of the whites who thirsts for our tears and listen to the voice of liberty that speaks in the hearts of all of us." The conspirators then took an oath of secrecy and revenge, an oath sealed by drinking the blood of a black pig they offered in sacrifice. The revolt had begun.

Setting fire to the sugar fields, the rebel slaves burned and tortured their former oppressors. In the first eight days of their insurrection, they destroyed nearly 200 sugar plantations. By the end of September, the slave army numbered between 20,000 and 80,000. "There is a motor that powers them and keeps powering them and that we cannot come to know," wrote one planter who had only narrowly escaped death.

They did not know it yet, but these slaves had initiated one of the most radical revolutions in the history of the Atlantic world. Over the next twelve years, these rebels fought and defeated the local white planters, the soldiers of the French empire, a Spanish invasion, and a British expedition of 60,000 men. But their greatest challenge would be the mighty armies of the French emperor, Napoleon Bonaparte.

In control of France by 1800, the great conqueror of Europe was plotting the creation of a "Republic in the New World," with Saint Domingue at the center and the North American colony of Louisiana as the breadbasket for the sugar island. In 1800, he ordered Charles Victor Emmanuel LeClerc, his right-hand man and brother-in-law, to subdue Saint Domingue, backed by a force of 42,000 battle-hardened men. These were troops that had defeated the most powerful armies of Europe: Austria, Prussia, Spain, Belgium, Italy, and the Netherlands.

LeClerc landed in Saint Domingue expecting easy victory. And in the first few months, he obtained it. Within ten days, the French controlled most of the island's ports and cities. Within three months, the French controlled nearly the entire island and had forced the main Haitian generals—including former slave turned commander of Haiti Toussaint L'Ouverture—to lay down arms.

But the rebels did not give up. French soldiers marched out into the countryside and the slaves melted into the hills, holding out in hopes of outlasting the invading force. Fate came to their aid. Yellow fever was ravaging the French army. And though the French now controlled the island, almost half of their military force died of disease. By the end of 1802, LeClerc himself fell prey to the dread disease. His second-in-command, Rochambeau, took over in his place.

Before he died, LeClerc declared that Saint Domingue could only be secured through a "war of extermination." He believed he would simply have to kill any black person who had ever been involved with the rebellion.

Thus began perhaps France's darkest hour. In desperation, in 1802 Rochambeau brought in packs of bloodhounds trained in Cuba to eat human flesh and unleashed them on the battlefield. But the dogs were "ignorant of color prejudice" and ate French soldiers as well. Rochambeau ordered slaves burned alive, drowned in sacks, or shot after digging their own graves. He became legendary for his brutality. But the slaves did not surrender, and by November of 1803 the rebel forces had driven what remained of Napoleon's soldiers out of the country. Over 80 percent of the French army sent there died on the island.

Amid the blood and destruction, Jean-Jacques Dessalines, the leader of the revolt and Toussaint L'Ouverture's successor, proclaimed the eternal freedom of the Haitian republic. "Let us imitate those people who, extending their concern into the future and dreading to leave an example of cowardice for posterity, preferred to be exterminated rather than lose their place as one of the world's free peoples," he declared. Victorious, black Haitians abolished slavery, declared racism illegal, and fought the first successful anti-imperial revolution in the history of the Atlantic. They also forever banned

Frenchmen from the colony. "May the French tremble when they approach our coasts, if not by the memory of the cruelty that they have inflicted, at least by the terrible resolution that we are about to take to devote to death, anyone born French, who would dirty with his sacrilegious foot the territory of liberty," Dessalines said.

The slave-rebels had beaten back the most powerful armies in Europe, overturned the prime economic engine of Enlightenment Europe, and struck the first victory in the war against slavery. And the vast Atlantic world of ships and slaves, of commerce and capital, could not help but take notice. In 1789, Saint Domingue exported 70,000 tons of sugar: by 1801, it exported only 9,000.

* * *

News traveled fast. Upon hearing of his brother-in-law's defeat in Haiti, Napoleon pounded the table and cursed, "Damn sugar, damn coffee, damn colonies." A strategic mastermind, the emperor knew when to cut his losses—as well as where to focus his energies. While a republic in the New World would have been nice, Napoleon had to focus on Europe and simply could not afford the massive costs in men and pride of subduing Haiti or running sugar colonies in the New World. With Haiti in flames, he saw little use for his other New World colony, Louisiana.

A headache to Napoleon, Louisiana was the apple of young America's eyes. Louisiana had a strategic place in the North American continent: its capital, New Orleans, controlled the Mississippi River. With a valley double the size of the Egyptian Nile and a drainage basin only slightly smaller than the Amazon, the mighty river loomed as the central artery of the American heartland, embracing 41 percent of the North American continent in its watershed.

As American settlers crossed the Appalachians and began to domesticate the West in the wake of their own successful revolution against the British from 1776 to 1783, they needed an outlet for their goods. The Mississippi River provided the only real channel for moving crops from the center of the continent out into the ocean and around back to the East Coast or to Europe; crossing the mountains by land was too great an obstacle. And just as the Mississippi River was the key to trans-Appalachian commerce, Louisiana—and New Orleans in specific—was the key to the Mississippi River. Guarding the outlet into the Gulf of Mexico, New Orleans was the single most important strategic site in North America west of the Appalachians. And Thomas Jefferson and his fellow republicans knew it.

But the demise of Saint Domingue and the rise of a free Haiti had wrought radical change on Louisiana society. Within a few short years, slave-rebels had sent the most profitable produce of the French empire up in smoke. Planters in Louisiana, at the time a military outpost surrounded by cotton, indigo, and sugar plantations, saw an opportunity for profit and rapidly began converting their fields for sugar production. An influx of Haitian refugees only added to the momentum. By 1802, a mere seven years after the first planter converted his entire plantation to sugar, Louisiana boasted seventy sugar plantations producing over 3,000 tons of sugar per year.

While Louisiana's yield still paled in comparison to what Haiti had produced in its prime, these numbers were enough to attract merchants from all over the eastern seaboard. By the turn of the century, sugar was becoming an increasingly common part of everyday life and demand was soaring. As Americans and Europeans drank more tea, smeared more syrup on their bread, baked more sweet cakes, and mixed more puddings and porridges, they needed more sources of raw sugar. Ships began to flock to New Orleans, where they filled their holds with what was fast becoming a staple of working-class diets. In a few short years after the slaves of Saint Domingue took up arms and formed themselves into a vast army, Louisiana was transformed from a small military outpost with a diverse agricultural mix into the center of the North American plantation world, one that revolved around sugar.

Haiti had affected not merely the world of European diplomacy, but the vast underworld of sailors, slaves, and debtors that made up the Atlantic underclass. Stories of the revolution, violent political ideals, and a commitment to freedom at all costs were spreading like a contagion from person to person—creating an epidemic that the planters of Louisiana could barely begin to understand. To the planters, the Haitian rebels were like rabid dogs. They saw insanity and bloodlust, rather than any political vision or humanistic ideal.

As aristocratic French planters like Jean Noël Destrehan worked to build a new Saint Domingue on the shores of the Mississippi, they did not realize the extent to which they were also creating the conditions that allowed the Haitian revolution to occur. More than any other place in North America,

Louisiana was becoming known for its brutal conditions. When slaves across the United States spoke with dread of being "sold south" or "sold down the river," they were speaking of the slave plantations around New Orleans. Nowhere in America was slavery as exploitative, or were profits as high, as in the cane fields of Louisiana. Slaves worked longer hours, faced more brutal punishments, and lived shorter lives than any other slave society in North America.

But as planters and government officials raked in the profits from this exploitative situation, they could not quiet the revolution the black Haitians had unleashed. Neither the American immigrants who rushed into Louisiana nor the long-settled French planters they met there fully realized the dangers that threatened the new order they hoped to establish. Unbeknownst to them, the slaves who labored on the region's sugar plantations were preparing to stage the greatest challenge to slave power in the history of North America.

Four

EMPIRE'S EMISSARY

In 1803, keenly observant of Napoleon's preoccupied state, Jefferson sent a representative, Robert Livingston, to negotiate for the acquisition of New Orleans and its environs. Despite his discomfort with purchasing a colony, Jefferson authorized Livingston to pay up to $10 million for the city—believing the acquisition of the port essential to national security. Upon hearing of Livingston's offer, Napoleon saw a chance to finally get rid of his troublesome American colonies and to make some money to fund his European wars at the same time. He offered to sell the United States all of Louisiana for only $15 million in cash. Without waiting for Jefferson's approval, after just nineteen days of negotiation, Livingston accepted the offer on behalf of his nation.

It was a massive purchase at a bargain price. The new territory doubled the young republic's size. Jefferson's $15 million bought what comprises about a quarter of the current geography of the United States—all of present-day Arkansas, Missouri, Iowa, Oklahoma, Kansas, Nebraska, parts of

Minnesota, most of North Dakota, nearly all of South Dakota, northeastern New Mexico, and portions of Montana, Wyoming, and Colorado. It was a diplomatic coup of gigantic proportions, a significance not lost on Jefferson and his contemporaries. "We have lived long, but this is the noblest work of our whole lives," said Livingston. "From this day the United States take their place among the powers of the first rank."

But what the Americans did not realize was just how foreign Louisiana was—and the host of difficulties they would face in taking control over this strange new land. Colonized by the French and controlled at times by the Spanish, Louisiana was more Caribbean than American—a place more similar to Haiti than to Virginia.

The boom in sugar plantations and the need to administer what had become a real slave colony made Louisiana even more problematic for an American government inexperienced with the problems of empire. Jefferson and his government quickly demonstrated the degree to which they underestimated the difficulties of governing Louisiana. To administer this vast area, Jefferson turned to William C. C. Claiborne, a fellow Virginian and political disciple distinctly lacking in qualifications.

Not even Jefferson thought first of Claiborne as governor. When he acquired Louisiana from Napoleon in 1803, he had first sought out the Marquis de Lafayette and then James Monroe. After both declined, he turned to William Claiborne—a "secondary character" whom Jefferson appointed at first on an interim basis. Claiborne arrived in New Orleans with a force of 350 volunteers and eighteen boats—a "puny force" that his top general described as "a subject for ridicule."

Claiborne had his work cut out for him. Only about 10 percent of the residents of New Orleans were Anglo-American; the rest were French, Spanish, African, Native American, or Creole (a person of foreign ancestry born in the New World). These residents did not look fondly upon Anglo-American outsiders like Claiborne. "The prejudices of these newly acquired citizens [are] against every thing American," wrote a correspondent to the *Orleans Gazette for the Country*. Yet American it was, a new national territory devoted to a single, slave-made staple crop.

When the United States took power in 1804, Claiborne spoke and advised the residents of New Orleans to "guide the rising generation in the paths of republican economy and virtue." He imagined he could transform this land into a new Virginia. He believed the power of the principles of self-government would naturally create a governable republic. But the simplicity of his scheme did not match the complexities of this wild city, with its proud and autonomous French planters, its anarchistic borderlands, and its dark and mysterious underworld of African slaves.

Looking back, such dreams might have seemed ignorant at best and arrogant at worst, but Claiborne's beliefs fell into neither category. Claiborne was a romantic in love with the ideology of the American Revolution. His father was a veteran of the American Revolution, and Claiborne had early internalized his father's love of country.

Growing up in Virginia, young Billy Claiborne (as he was

known to his family and friends) used to listen raptly as his father spoke in "glowing colours" against the horrors and brutality of the British. Colonel Claiborne would rail about the "do nothings," the "armful of sulking slackers who cowered on the side-lines and cheered whichever team seemed to be winning." For the colonel, the foundation of liberty and the creation of the American Republic were the greatest and proudest moments of his life. And he cast constant denunciations on anyone who might dare to "raise a parricidal hand to destroy the fair fabric of American liberty." Evidence suggests Billy internalized these early lessons. When only eight years old, he turned in a Latin composition that read, "Dear my country, dearer liberty—where liberty is, there is my country." Young Claiborne did not consider that the massive slave population might feel the same way; his conception of liberty extended only to white males.

Moreover, perhaps ironically, Claiborne believed liberty could be imposed from above. Like Thomas Jefferson, he saw Louisiana as an imperial colony of alien people who needed to be Americanized with a firm hand. Claiborne wanted the new territory of Louisiana to become American, not merely be an American colony with a French culture. Claiborne had little regard for Europe in general, or France in specific. He dismissed the "corrupt governments of Europe" and expressed no interest in learning European languages. He was a son of Virginia and that was where his heart lay. "The very trees that had shaded him from a summer's heat, were with him objects of veneration," wrote Claiborne's brother. He worshipped the "everlasting marble records the names of the first proprietors." He was, to say the least, an unlikely ambassador to the proud Frenchmen of New Orleans.

In the first decade of American occupation, Claiborne had to form a government, bring order to a wild frontier zone, and confront the dangers of a sugar colony that relied on the forced labor of a slave population. New Orleans was the most diverse, cosmopolitan, and European city of North America, but Claiborne intended to rapidly make it American. Jefferson's initial plan was to pay for 30,000 Americans to immigrate into the new territory and "amalgamate" with the French residents. "This would not sweeten the pill to the French," Jefferson wrote, "but in making that acquisition we had some view to our own good as well as theirs." Governor William Claiborne, who spoke neither French nor Spanish, would be in charge of this grand task, assisted by a top general, the questionably loyal James Wilkinson.

New Orleans society did not look favorably on the newcomer's attempts to instill American values in a much older and longer-established French society. "All Louisianians are Frenchmen at heart!" wrote one French official. The French Creoles formed an aristocracy of the blood, impenetrable to outsiders and marked by snobbery. Tracing their ancestries back to French nobility, the planters condemned lesser families as *chacas, catchoupines, catchumas,* and *kaintucks*—referring in order of social status to tradesmen, peasants, people with African blood, and Americans.

The planters were more interested in parties than in the blessings of republican self-government. When Claiborne arrived in January 1804, the French planters informed him that a celebration was absolutely necessary to win their support and ensure American control. Some 196 gallons of Madeira, 144 bottles of Champagne, 100 bottles of "hermitage" wine, 67 bottles of brandy, 81 bottles of porter, 258 bottles of ale, and 11,360 "Spanish Segars" later, an exasperated Claiborne offended the entire French population by publicly declaring that the French planters would never understand what it meant to be American. In a letter to President James Madison, Claiborne wrote that the greatest of the planters' "mischiefs" was moral depravity. Their love of money, luxury, and debauchery "had nearly acquired the ascendancy over every other passion."

Destrehan and his social circle soon taught Claiborne the consequences of interfering with their long-established culture. At a dance that same year, a self-righteous Claiborne ordered the assembly to dance an English dance before the French dances began. The French planters began to raise a hullabaloo, shouting and carrying on. General Wilkinson attempted to address the planters in broken French, but that only made matters worse. To drown out the uproar, Claiborne and the American officers with him started singing "Hail Columbia," but the Creoles responded with "La Réveil du peuple" and shouts of "*Vive la République*." Tension soon bubbled into open brawls. Fearful of what might happen next, Claiborne and some of his officers beat a hasty retreat out the back door. Claiborne wrote that from the balls "have proceeded the greatest embarrassments which have hitherto attended my administration."

After this catastrophic event, recriminations flew in the local newspapers. "Does [Claiborne] think he is among Indians or Yahoos?" wrote one newspaper columnist, accusing Claiborne of being an "uncouth and ignorant intruder" into New Orleans society. Another columnist attacked the American governor for his inability to speak French, his unfamiliarity with French dances, for being embarrassed by "the insignificant part he acted in the circle," and for sneaking home at sunrise after losing at the gambling tables. Claiborne, in turn, believed the French planters were unfit for self-government.

The conflicts in the New Orleans social world soon spilled out into the larger political sphere. In 1805, Jean Noël Destrehan had led a delegation to Washington, D.C., to protest Claiborne's appointment and the "oppressive and degrading" form of the territorial government. The delegates bemoaned the "calumnies which represent us in a state of degradation, unfit to receive the boon of freedom," demanding immediate citizenship and statehood. Destrehan deeply resented the Americans' treatment of the French planter class—and especially Claiborne's portrayal of the planters as unfit for self-government. "To deprive us of our right of election, we have been represented as too ignorant to exercise it with wisdom, and too turbulent to enjoy it with safety," he wrote.

Attacking the arrogant officials who sought to govern Louisiana, "who neither associate with us, nor speak our language," Destrehan and his friends let off a targeted attack on the monolingual Claiborne. Destrehan wrote that the Spanish were "always careful, in the selection of officers, to find men who possessed our own language, and with whom we could personally communicate." Rather than study French,

Claiborne preferred spending time with older women, "to whose conversation and company through life he was most passionately devoted," as his brother wrote.

Destrehan touted the great abilities and virtues of the planters. He focused, unsurprisingly, on their noble lineage. "We were among the first settlers; and, perhaps, there would be no vanity in asserting that the first establishment of Louisiana might vie with that of any other in America for the respectability and information of those who composed it." Destrehan saw this as an infallibly good argument.

Destrehan laid out a clear vision for two possible futures: one marked by continuing tensions with the U.S. government and one marked by recognition of the French planters as citizens and as sovereign people capable of self-government. "Annexed to your country by the course of political events, it depends upon you to determine whether we shall pay the cold homage of reluctant subjects, or render the free allegiance of citizens," Destrehan wrote.

Claiborne and the government in Washington chose not to honor Destrehan's requests. Expressing deep doubts about the honesty and trustworthiness of the planters, Claiborne encouraged Madison to give them nothing. "The people had been taught to expect greater privileges, and many are disappointed," he wrote. "I believe, however, as much is given them as they can manage with discretion, or as they ought to be trusted with until the limits of the ceded territory are acknowledged, the national attachments of our new brothers less wavering, and the views and characters of some influential men here better ascertained." He expressed particular doubt about Destrehan and his friends, and worried that allowing the citizens of Louisiana a representative rather than imperial government would be "a hazardous experiment." Henry Adams, a prominent historian of the period, wrote, "the lowest Indian tribes had more right of self-government than members of Congress are willing to give the people of lower Louisiana."

Claiborne complained constantly of New Orleans's diverse mix of Spaniards, Frenchmen, and African slaves. "Renegadoes from the Atlantic states, who repairing in shoals to New Orleans, more greedy than the locusts of Egypt, expecting and soliciting all the offices in the gift of the new government, and when disappointed, setting up and supporting venal and corrupt presses to vilify and abuse him, and to exhibit in an odious point of view every act of his public life that envy and malice could seize on as the subject of accusation," Claiborne's brother, Nathaniel, wrote, describing the hardships that William complained of in his first years.

Claiborne was right to complain of the planters' printing presses, which they used on many occasions to attack him personally and politically. Perhaps their worst moment of spite came soon after the death of Claiborne's wife Eliza from yellow fever. In popular newspapers, they portrayed the governor's social life as a constant attempt to marry up—to find a richer and more socially prominent French woman who would aid him in his quest for power and acceptance. They spread rumors that he wanted a woman who would help him overcome his "pecuniary difficulty."

A newspaper satirist wrote about a dream in which he was walking through the quiet and dark streets of New Orleans late one evening. But turning a corner, he came upon the governor's mansion, where he heard music and dancing and saw bright lights from the windows. While gazing up, he caught sight of the ghost of Claiborne's dead wife gazing up too at her husband's party. Turning away in agony at the thought of her husband celebrating so soon after her death, she "bent her willing steps towards the graves of Louisiana." One can only imagine how Claiborne felt upon reading this particular attack—or experiencing the general vitriol of the arrogant Frenchmen he intended to govern. Bursting with high ambitions, Claiborne was highly sensitive to criticism and took great offense at these attacks.

Put upon and attacked—even dismissed—by the planters and their society, Claiborne labored on, seeking to introduce the principles of liberty and republican self-governance to these decadent Europeans. Lost in their disputes over dances and languages, the white elite seemed to have lost sight of the larger problem with their frontier society. Squabbling over petty matters, they did not address the tremendous danger posed by the rapidly growing slave population. Focused on political intrigues, they did not notice the increasingly radical tenor of the political discussions in the slave quarters. They failed to realize that the true conflict at the heart of New Orleans was not between the French and the Americans but between the white elite and the vast African underclass.

Five

CONQUERING THE FRONTIER

WE SHOULD HAVE SUCH AN EMPIRE FOR LIBERTY AS SHE HAS NEVER SURVEYED SINCE THE CREATION: & I AM PERSUADED NO CONSTITUTION WAS EVER BEFORE SO WELL CALCULATED AS OURS FOR EXTENSIVE EMPIRE & SELF GOVERNMENT.

Thomas Jefferson

In the months leading up to January 1811, the troublesome French planters were not the greatest of Claiborne's problems. Claiborne had focused all of his energy and attention on resolving a grave threat to national security.

While New Orleans was under American control, the city was surrounded by Spanish territory. The Spanish controlled an empire in the Americas that extended from Florida in the east to the Pacific coasts of Mexico and California in the west. Most significantly to Claiborne, however, they controlled the region known as West Florida, which extended from Baton

Rouge on the southwest and Natchez on the northwest to Mobile on the east.

The threat to national security came not from the Spanish army, but from the fragile state of the Spanish government. Napoleon had recently conquered Iberian Spain and placed his brother Joseph on the throne—a move opposed by popular juntas that refused to declare allegiance to the "intruder king." The civil unrest in Spain left the colonies in disarray, with power devolving to local officials and the small garrisons they maintained. With no central government, the Spanish colonies had descended rapidly into a state of near-anarchy.

To prop up their collapsing empire, the Spanish governors allowed native tribes, escaped slaves, and profit-seeking farmers to settle on Spanish land—provided they swore an oath of allegiance to the Spanish crown. The Spanish encouraged and sponsored Native Americans, free blacks, and runaway slaves, attempting to convert these outcasts of American society into allies and supporters of the Spanish cause. The Spanish lands provided a beacon of hope for slaves seeking to escape and rebel, as well as a safe haven for Native American exiles who had been forced off their land by the expansion of American agriculture. Claiborne feared that this chaos could spread like a cancer, corrupting the order he sought to bring to New Orleans.

In 1810, Claiborne decided to take decisive action. He traveled to Washington, D.C., to secure approval for a covert paramilitary action to topple the Spanish government in West Florida without starting a war between the United States and Spain. Because open conquest was blatantly illegal, Claiborne knew that America could not be seen to have instigated this plot. In Washington, Claiborne presented his plan to President Madison: a small group of handpicked adventurers would be enlisted to attack the Spanish garrison, seize power, declare independence, and then request annexation to the United States. The United States would condemn the action publicly and reluctantly agree to annex the rogue state for the sake of national security.

Claiborne and Madison both saw the strategic advantages of such an operation. Conquering Baton Rouge and West Florida would give America full control over the Mississippi River, eliminating a haven for escaped slaves and dangerous native tribes while securing commerce on the river. As an added benefit, West Florida was full of Anglo-American settlers, and Claiborne knew that if he could include West Florida in the Orleans Territory, he could tip the balance of power toward American authority and undermine the French planters' still-strong political position. After carefully weighing the pros and cons, Madison gave Claiborne approval.

With Madison's approval, Claiborne drafted a letter to William Wykoff Jr., a wealthy plantation owner and member of the Orleans Territory executive council. Wykoff was well-connected with the Anglo-American settlers in West Florida and seemed to Claiborne an ideal agent. In the letter, Claiborne sketched out the difficulties facing the Spanish government: the Napoleonic conquest, the unrest in Latin America, and America's sketchy legal claim to much of the land in West Florida under the terms of the Louisiana Purchase.

The American military would take possession of West Florida, Claiborne hinted, but not as a straightforward military conquest. "It would be more pleasing that the taking possession of the Country, be preceded by a Request from the Inhabitants," Claiborne wrote. "Can no means be devised to *obtain such a request?*" Claiborne suggested a set of rich and prominent planters in the Baton Rouge area as potential supporters for this effort, and he assured Wykoff of the "friendly disposition of the American Government" to any "decided measures" these men might take. Claiborne stopped short of saying what exactly these measures might be, but simply suggested that Wykoff "lose no time in sounding the *views* of the most influential of your Neighbors on the opposite Shores, and in giving to *them* a right direction." In closing the letter, Claiborne instructed Wykoff to keep the letter absolutely confidential, and to feel free to leave off his signature in future letters to Claiborne.

In the summer of 1810, Wykoff rode to the doorstep of a new settler in the region, Fulwar Skipwith. "Endowed with more than average intelligence, well cultivated by collegiate study, and by his Cosmopolitan associations," Skipwith "was more than six feet tall, straight as an arrow, with exactly enough flesh for his bone and muscle," in the words of one friend. Wykoff knew he could trust Skipwith's loyalty to the American government—he was an eighth cousin of Thomas Jefferson. Using money from his recent marriage to a Flemish baroness, Skipwith had built a 1,300-acre plantation called Monte Sano along the Mississippi River south of Baton Rouge in 1809. He was eager to earn a name for himself—and to support the efforts of Claiborne, a fellow Virginian. He had no problems understanding Wykoff's plan. He understood, he wrote, that it would be "more satisfactory" for a convention of "honest cultivators of the soil" to overthrow Spanish authority and request annexation by the United States than for the American army to fight the Spanish openly. The device would allow the United States government to avoid violating international law while still obtaining possession of the desired territory. "All would be washed, except the poor Floridian, in holy water," Skipwith wrote, and "sweet . . . to the palate of the . . . administration."

Before dawn on September 23, 1810, Skipwith appeared outside the Spanish fort at Baton Rouge with eighty armed American settlers. The fort was situated on high ground overlooking the Mississippi River, surrounded by cypress pickets that slanted outward to protect against invaders. Banks of clay as high as the pickets formed walls that ran between bastions at each of the four corners of the fort. A dry ditch around the outside of the fort added to the defenses. The fort was well guarded: four cannons threatened at the main gate, blockhouses dotted with musket portals, and a band of twenty-eight Spanish soldiers under the command of a young lieutenant named Louis de Grand Pré. The Spaniards were confident of their abilities to beat off any invading force; they did not expect what came next.

A young Kentuckian named Larry Moore had informed Skipwith that he knew how to "get inter the dinged ol' fort," describing a small opening in the cypress palisades by the river where the Spanish brought in cows for fresh milk. "Ef them cows kin get in thar an' outen again, I knows my pony kin tote me in the same way, an' do h'it as easy as fallin' offen a log," said Moore, spitting out a wad of chewing tobacco. Following Moore, Skipwith's horsemen circled the fort and guided their horses through a herd of feeding milk cows. Almost immediately, they penetrated the fort and entered the center of the garrison. Emerging into the open, the Americans shouted, "Ground your arms and you shall not be hurt." But Grand Pré and his men refused to surrender. Standing by the Spanish flag, they fought desperately against the American attackers in hand-to-hand combat. But the Americans had the advantage. "Shoot 'em down," shouted Isaac Johnson, Skipwith's deputy. Hails of lead quickly disposed of several Spanish soldiers and the Americans whooped "Hurrah Washington!" at the top of their lungs. One American smashed his rifle butt into the head of the Spanish governor. By midday, the Americans were in full control of Baton Rouge.

Skipwith proudly declared the independence of the new Republic of West Florida. The Spanish flag disappeared and the Bonnie Blue Flag with a white star fluttered in the wind over the fort of Baton Rouge. "Betrayed by a magistrate whose duty was to have provided for the safety and tranquility of the people . . . and exposed to all the evils of a state of anarchy," Skipwith declared, "it becomes our duty to provide for our own security, as a free and independent State." In the declaration that he released to the international community, Skipwith made no mention of ties to the U.S. government—or of any intention to ask for annexation. But through private channels, Skipwith made clear that his declaration was a smokescreen—and that his true loyalties lay with Madison and Claiborne. Skipwith expected to be greeted with congratulations and public celebration, but Claiborne had other plans.

Claiborne, who had stayed in Washington to avoid being implicated in the illegal conquest, rushed back to New Orleans in December. Publicly decrying "the intrigues of certain individuals . . . of desperate character and fortunes," Claiborne ordered the U.S. military to seize control of Baton Rouge and the other parts of West Florida conquered by Skipwith. His plan had succeeded.

Upon return to New Orleans, Claiborne set out to Baton Rouge to assert American authority there. Sending a friend to speak to Skipwith in advance, Claiborne proceeded up the Mississippi River with gunboats, dragoons, and infantry. After receiving word that Skipwith accepted the transfer of authority, Claiborne congratulated Skipwith on his "correct" conduct and assured him that no legal repercussions would follow for his actions. Unlike the more famed filibuster, former vice president Aaron Burr, Skipwith would receive no indictment for treason for his work to take over Florida. Assured that all was in order, Claiborne landed triumphantly on the right bank of the Mississippi River, once the territory of Spain and now fully American. The cavalry and riflemen received Claiborne on the beach, and he marched with pomp and circumstance to the pavilion where the flag of the Republic of West Florida flew. Claiborne read a quick proclamation, declaring that the United States would "protect them in the enjoyment of their liberty, property and religion," and ordered the American flag raised over Baton Rouge. On December 7, 1810, West Florida formally became a part of America.

Ten days later, Claiborne sat down to write a letter to the Spanish governor and captain general in Cuba to inform him of what had recently occurred in West Florida. Claiborne disavowed all affiliation with "the association of Individuals" who had established "an *independent state*" in West Florida. The actions of these men, Claiborne wrote, "gave to my *Government* much solicitude and imposed upon *it* the necessity of resorting to the most prompt and effectual means." Blatantly hiding the American association with these adventures and suggesting the United States government was forced by circumstance to annex West Florida, Claiborne sought to deceive the Spanish and prevent an international conflict. "Your Excellency will *not* I am persuaded, attribute this measure to an unfriendly disposition towards Spain," Claiborne wrote innocently. Claiborne portrayed the Americans as innocent respondents to a crisis—as having had no role in the course of events that led up to the attack on Baton Rouge. By feigning passivity, Claiborne promulgated a story about West Florida that has largely entered the history books—a story of a minor and inevitable border conflict in which the United States had no role in violating international law in order to conquer Spanish territory.

Claiborne believed strongly that American expansion was God's work, and that whatever actions he took to promote her power and improve her national security would ultimately prove justified by the blessings that would flow from enlightened government and individual liberty. He dismissed the laws of nations and the rules of European diplomacy as corrupt remnants of an old world, irrelevant in the face of the new order he was helping to create. When he spoke of Spanish control over West Florida, he spoke of it as a relic of history, as a footnote to a more important story about the rise of the United States. It seemed Claiborne would do almost anything to contribute to that grand narrative of American expansion, hushing up anything that ran counter to his grand vision of the republic.

* * *

But though America now officially controlled West Florida, Claiborne faced difficult problems the day he scrawled the date *January 6, 1811* on the top of a sheet of paper. West Florida remained in a difficult spot. Groups of armed American settlers roamed the territory, and Claiborne was as yet unsure if they intended to comply with American authority or would continue advocating for a Republic of West Florida. Some of these armed men were committing acts of piracy, stopping travelers on the Pascagoula River and demanding that they swear allegiance to Florida or else forfeit all of their property. One traveler reported the territory in a state of "absolute anarchy." A gentleman from Pensacola informed Claiborne that the Spanish were en route with 1,500 men and $500,000 from Havana to Florida to retake the territory. Reports were also arriving of a "terrible revolution" in the Kingdom of Mexico, where "ten thousand creoles" were reported to have been slaughtered. Finally, on the night of the fifth, a group of sailors started a riot in the harbor, perhaps in anticipation of Epiphany and the start of Carnival. Claiborne's mind was consumed with the question of how to combat the Spanish and assert full American control in the anarchical state of West Florida. He believed that a secure West Florida meant a secure Louisiana—and that a secure Louisiana represented the first step in building an American empire that would stretch from sea to shining sea.

As Claiborne ordered the majority of the troops under his command toward Baton Rouge and West Florida, he did not even consider the defenseless state of New Orleans. As far as he was concerned, Spain—and the anarchy in West Florida—posed the main threat to American authority in the territory. Caught up in his expansionist agenda, he did not even notice the unusual beat of the African drums in the marketplace or suspect the threat bubbling up beneath the placid surface of the plantation region outside the city.

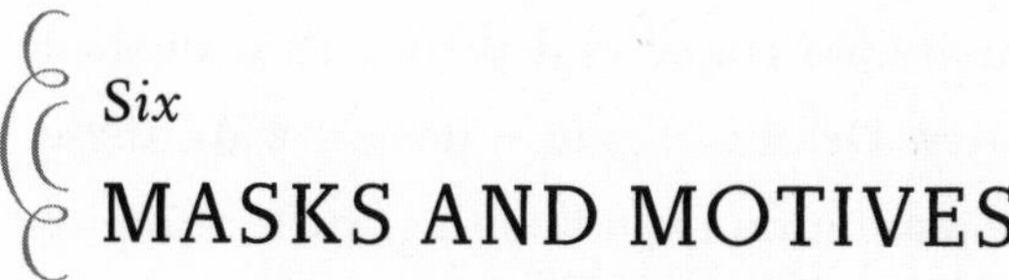

Six
MASKS AND MOTIVES

FOR IT IS THE SAME WITH THE BLACK AS WITH THE WHITE MAN. ASSEMBLE TOGETHER FOR THE FIRST TIME TWENTY OR FIFTEEN . . . MEN . . . AND WITHIN FORTY-EIGHT HOURS AFTER BEING BROUGHT TOGETHER, THOUGH STRANGERS TO EACH OTHER, THE GREAT MAJORITY WILL PLACE THEIR EYES ON CERTAIN MEN AMONG THEM, FOR THEIR WISDOM, COURAGE, AND VIRTUE, TO WHOM THEY, UNKNOWINGLY TO ONE ANOTHER, DETERMINE TO LOOK UP, AS LEADERS OR CHIEFS, TO CONDUCT, COUNSEL, AND ADVISE THEM.

Louisiana sugar planter John McDonogh

Forty-one miles northwest of New Orleans, set back a short distance from the Mississippi River, behind a field of clover, stood Manuel Andry's elegant two-story-high mansion. The house boasted a colonnaded porch and broad galleries that wrapped around the main structure. Large curtains extended from the pillars to defend the house from the heat of the summer sun. The front of the house looked out along manicured gardens and oaks dangling with Spanish moss toward a small ridge, four to six feet high, that ran along the near horizon.

This small ridge was the levee, thirty to forty yards beyond which the wild Mississippi roiled and turned as it flushed downward toward New Orleans. Six to nine feet wide, these levees protected the plantation from flooding, while also providing a convenient footpath for travelers and traders. Docks situated on the river connected the plantation to the river's transportation systems.

The fields ran from the levee along the river to the swamps, where another levee had been constructed. Property lines extended from the levee back into the dark swamps. Irrigation ditches and paths divided up the fields into gridded rectangles. The Mississippi River fertilized the rich soil mix of clay, sand, and vegetable mold. From this land grew the riches that allowed the rest of this world to be built: the sugar cane that was like white gold.

The planters structured their lands as sugar factories, creating soil and irrigation patterns suitable for the plants to grow and field patterns suitable for the proper supervision of enslaved laborers. Though an uneducated visitor might simply see fields and trees and structures, each indent in the land, each row of the cane, each building was positioned with skill and art to turn this land into a sugar-producing factory. And it was in these strange lands that the planters and the slaves interacted on a daily basis.

As was standard on German Coast sugar plantations, the

main plantation house stood by the river, with a road leading back to the sugar house near the back of the plantation. The slave quarters were positioned along the road. The slaves lived in small two-room brick cabins with a central fireplace. Each house held an entire family. The parents slept in the main room, while the children climbed into the attic. The brick residences were drafty and cool. While the plantation owners ate five-course meals, the forty or fifty slaves on each plantation ate stew or jambalaya, just enough to survive and work.

The planters planned every part of their plantations to harness land and labor in the most efficient manner. The very landscape of the plantation functioned to enable and maximize the efficacy of labor. The geometric rows and rectangles of the fields, similar to a military camp, enabled the master to see everything that went on in the plantation. A former slave from the region described this system of control. The overseer, "whether actually in the field or not, had his eyes pretty generally upon us," wrote Solomon Northup, a free Northerner who was kidnapped and sold into slavery in Louisiana. "From the piazza, from behind some adjacent tree, or other concealed point of observation, he was perpetually on the watch." By keeping constant watch over their slaves, the masters asserted control over their actions.

The land between the river and the swamps was the domain of the planter—the central zone of power and profit. But the master could not watch and manage every daily task and activity. Rather, he depended on slaves to do the watching, the supervision, and the daily direction of labor on the plantation. The system of slavery rested on coercing or bribing a portion of slaves into betraying their compatriots and becoming loyal tools of the planter elite.

Charles Deslondes was one of these slaves. A light-skinned black man, he served as a slave driver, a member of the slave elite. Slave drivers were a notoriously conservative group with a bad reputation as traitors to the slave cause. "It is fair to say that these slaves, intermediary links, who would fasten in some way the chain of servitude to that of despotism and would find a malicious pleasure in overburdening others with work and vexation, combined the baseness of their condition with the insolence of their authority," wrote the novelist Victor Hugo in a novel about a revolt on a sugar plantation. Drivers were infamous for their roles in whipping other slaves; some were even thought of as "human bloodhounds." In some cases, drivers would participate in the hunts for fugitive slaves, following the dogs that pursued those who ran away into the swamps.

Charles Deslondes had quickly risen through the ranks, driven by ambition, success, and a light skin tone that made him seem more trustworthy than the many Africans imported into the area. Born on the Deslondes plantation on the German Coast, Charles served as the driver for the Spaniard Manuel Andry, a planter known for his cruelty toward his slaves. As driver, Charles served as Andry's right-hand man, running the day-to-day operations of the sugar plantation.

This was no easy job, and it required intelligence, strength, and an ability to command and control large numbers of men. The process of cultivating sugar cane is more complicated than most agricultural enterprises. It requires sophisticated organization, time discipline, and strict scheduling. Each stalk of cane bursts with thick, rich juice. Slaves had to grind and boil the chopped cane to obtain this juice—a complex process that required industrial expertise on a level with that of Northern factories. Slaves developed complex skill sets, and they learned to work under military-style discipline.

Charles supervised all of this on behalf of Andry. Charles rang the bell that woke his fellow slaves for work every day. Charles monitored each slave's performance, whipping or otherwise punishing the lazy and the slackers. He broke in the new African slaves, forcing them to adapt to the rough rhythms of daily work. He carried the keys to all doors—including those behind which unruly slaves were kept. He communicated daily with his master, sharing information, planning the work calendar, and discussing the sugar crop—that great dominant force in the lives of the masters and the slaves. He was the right-hand man of the master, with his feet in the worlds of whites and blacks. Day after day, year after year, he managed the complex rhythms of sugar growing and harvesting, gaining status and better treatment for himself and untold riches for his master.

The process of growing sugar cane started in the cold of January, when the slaves plowed the fields to open up furrows for the year's seed. By February, the planting was complete. The planters assigned the slaves to tend the crop, weeding and irrigating, and guarding against insects and other dangers. During the hot summer months, they turned their attention to the wide range of other plantation tasks: repairing levees, making bricks, mending roads and fences, growing provisions, gathering wood for fuel, and getting ready for the fall harvest.

This work was nothing, however, compared to the most trying and essential part of the crop cycle: the fall grinding season. During this season, the slaves raced against time to harvest the entire crop before the first frost. Planters delayed the harvest as long as possible because the longer the cane stayed in the ground, the richer and more valuable it became. Once the harvest began, the slaves worked sixteen or more hours per day, seven days a week.

To organize this labor, most planters divided the slaves into three gangs, each led by a slave who would report directly to a driver like Charles. The first gang, made up of the strongest and most powerful young men, used fifteen-inch-long knives to cut the mature eight-foot-tall cane. Sugar cane is essentially a tall grass, with sweet, juicy stalks that can grow as thick as two inches, and as high as fifteen feet. This gang proceeded down the rows of cane, with one slave leading the cutting, one cutting to the left, and the other to the right, depositing the sugar cane in the middle of the row. The second gang was made up mostly of younger slaves and women. These slaves loaded the cane into carts, hauled by mules, and took the cane to the sugar mill. The third gang, highly skilled laborers who knew the intricacies of cane sugar, ran the sugar mill. Working round the clock—feeding wood into the fires, watching the boiling kettles, and moving sugar through the process of granulation and purification—this last gang kept the mill going constantly from mid-October through Christmas and often into January. According to a French official, the system had been pioneered by the planter Jean Noël Destrehan, who "by a wise distribution of hours, doubled the work of forty to fifty workers without overworking any of them."

Slaves had to deal not only with hard work but also with a difficult natural environment. The heat in the summer months was unrelenting, and the swamps made the environment particularly dangerous. Many slaves fell prey to tropical disease. For much of the year, mosquitoes made being outdoors unbearable. "From June to the middle of October or beginning of November, their swarms are incredible," architect Benjamin Latrobe wrote in 1819. "The muskitoes are so important a body of enemies that they furnish a considerable part of the conversation of every day and of everybody; they regulate many family arrangements, they prescribe the employment and distribution of time, and most essentially affect the comforts and enjoyments of every individual in the country." Mosquitoes were not just pests; they were vectors of malaria and other tropical diseases, and they represented one of the greatest challenges to keeping slaves alive long enough to make a profit from their labor. Sugar plantations were deadly places.

Planters like Andry relied on drivers like Charles to listen to their slaves, to understand their problems, and to keep an ear open for discontent. More often than not, drivers fulfilled the function of diminishing tensions between whites and blacks, keeping the machinery of the slave plantation going. The Louisiana slave Solomon Northup recalled the frequency of talk of insurrection in the slave quarters—and his own conservatism. "More than once I have joined in serious consultation, when the subject has been discussed, and there have been times when a word from me would have placed hundreds of my fellow-bondsmen in an attitude of defiance without arms or ammunition, or even with them," he wrote. "I saw such a step would result in certain defeat, disaster, and death, and always raised my voice against it." Most drivers, it seemed, used their place on the middle ground to quell, rather than fan, the fires of rage and violence omnipresent in slave society.

They were also responsible for administering the brutally violent punishments that functioned to keep order on the plantation. A sugar plantation was like a military camp, and Charles was the general. The complexity and intensity of sugar farming demanded militaristic management styles. The simple goal of the driver was to turn these subordinate slaves into sugar-producing machines. "The feelings of humanity remain inert when it comes to the slaves," wrote a traveler passing through in 1803. "The purpose of slavery is only to tie down the blacks so that they work the land like oxes or mules. To insure this result, there exists an organized hierarchy of drivers, chiefs, and overseers, always whips in hand." Charles carried that whip.

A first punishment for a more minor transgression might be imprisonment. Charles had full authority to chain up or lock away any disobedient or discontented slave. Each plantation had a place to imprison, detain, or chain up recalcitrant slaves. Jean Noël Destrehan, for example, used the small washhouse behind his mansion as a dungeon for recalcitrant slaves.

But the next, and perhaps most common, punishment was the whip. Charles might lick the back of a slow-moving harvester, or take aside a slave who talked back to him. But for the more serious punishments, the whipping became a public demonstration of cruelty and power. One Louisiana overseer described the process:

> *Three stakes is drove into the ground in a triangular manner, about 6 feet apart. the culprit is told to lie down, (which they will do without a murmur), flat on the belly. the Arms is then extended out, side ways, and each hand tied to a stake hard and fast. The feet is both tied to the third stake, all stretched tight, the overseer, or driver then steps back 7, 8, or ten feet and with a raw hide whip about 7 feet long well plaited, fixed to a handle about 18 inches long, lays on with great force and address across the Buttocks, and if they please to assert themselves, they cut 7 or 8 inches long at every stroke.*

This inhuman torture reminded slaves of their inferior position—and their less-than-human status.

For some of the worst and most recalcitrant offenders, the French planters designed torture devices that would constantly remind the offender of his or her actions. Collars, metal masks, and other such devices were far from uncommon. One common device featured a neck collar with inward-pointing spikes that prevented the victim from lying down and resting his or her head.

Behind all of these punishments lay a force greater than the whip or the spiked collar, a force which ultimately provided the backing for the master's rule on the plantation—death. Violence and the threat of death were the essential elements of the commoditization and enslavement of people. However, death was a card slaveholders were reluctant to play, and slaves understood and knew that reluctance. The corrective forms of discipline were means of organizing labor and maximizing efficiency without recourse to that ultimate form of violence. Violent punishment and slave resistance were in constant dialogue—two sides of the same coin. The planters understood the danger that slave resistance posed to their livelihoods.

While many other slave societies in the United States were self-reproducing, no such calculus existed in Louisiana. Sugar work was too grueling and demanding, the profits too large, and replacement slaves too easily available to worry much about natural reproduction. In 1800, one planter estimated that each plantation hand produced $285 per year, with the average hand priced at $900. Within four years, a slave had more than recouped the initial investment, rendering the need for natural reproduction less important. Planters relied first on the Atlantic slave trade and then on the internal slave trade to supply a steady stream of new workers.

Amid these horrible conditions, no individual planter had the power to stop his slaves from revolting, not when they outnumbered him fifty to one. The planter relied on slave drivers, gang leaders, and bribed or coerced informants to maintain a militaristic choke hold on the people who labored for them in these plantations. Those who complied did not go unrewarded. Charles would have been compensated for his good work and loyalty with a larger hut, nicer clothes, or gifts of money or extra food.

Other than his role as a driver, most slaves knew only one thing about Charles: his relationship with a woman on the Trépagnier estate. Drivers like Charles were trusted to travel more freely than any other slaves, and Charles took advantage of his relative freedom to leave the plantation frequently, spending nights and weekends in a small cabin with a woman far from the Andry estate. Manuel Andry himself permitted these visits, and Trépagnier sanctioned them, perhaps in the hopes that Charles would impregnate this woman and she might bear a child as loyal and hardworking as his father. But no one really kept track of Charles's conjugal visits. Few knew much about whom he really spent time with or whom he confided in. He kept a tight circle of confidantes, but other than that he was an unknowable. To most slaves, he was simply the half-white representative of the master. And to the master, he was the half-white liaison from the slave quarters. He was the central link, the connector, and the enabler of the complex machinery of the Andry slave plantation—or so it seemed.

Seven
THE REBELS' PACT

As Charles walked the few miles from the Andry plantation to the Trépagnier estate to see his woman, he passed by the plantation of James Brown. Here he must have stopped to talk with Kook and Quamana before continuing on. Perhaps as he went to New Orleans on his master's business, he would linger beneath the shade of the cypress trees that lined the fields and speak with other slaves from other plantations.

Romance provided the perfect cover. While it was not a formal marriage, Charles had taken up with a slave on the Trépagnier plantation, whom he visited whenever he could. He did not marry this woman—whose name is lost to history—perhaps because he was not allowed to, or perhaps because he understood how unrealistic marriage was in a society where a master could rape or sell one's spouse at his own convenience. Slave women had little control over their reproductive and sexual lives, and they were the constant victims of rape and sexual violence by every white male, from the master and his sons to neighboring planters and itinerant laborers. Relationships between slaves were vulnerable to the sexual whims of the master class, and recognizing this, men like Charles often chose to simply take up with a woman rather than cement a relationship with easily broken marriage bonds. Charles's relationship with this woman was no doubt constrained by the tragic realities of master-slave power relationships.

As to his light skin, most everyone knew the source of that: he was the son of a white planter—a white planter who had slept with and impregnated Charles's slave mother. Such relationships were not uncommon. The abolitionist Wendell Phillips famously condemned the antebellum South as "one great brothel," where every slave woman lived in fear of coerced sexual activity, and where interracial sex was an obvious, though discreetly discussed, element of the landscape. And much as Charles might want to suppress thoughts of his own mother's violation, his paternity shaped every moment of his daily life and career on the German Coast. Charles's light skin differentiated him from the other slaves in the eyes of the planter class and of the slaves.

But as Charles went about his daily life and work, ringing the bells, whipping the slaves, driving ahead the machinery of the sugar factory, taking nights and weekends at the home of his mistress, he was not the contented slave he appeared.

In fact, he was using his authority, his relative freedom, not on behalf of his master but rather to push his own agenda. Charles, who in the eyes of the planters and his fellow slaves seemed to be the most loyal and the most privileged of all slaves, was in his spare time a plotter. He was one of the key architects of an elaborate scheme to kill off the white planters, seize power for the black slaves, and win his own freedom and that of all those laboring in chains on the German Coast. He was, in modern terminology, the ultimate "sleeper cell," imbedded intimately close to the enemy he dreamed nightly of executing. He would begin his revolution by attacking Manuel Andry.

We will never know what motivated this fateful decision, what factors Charles weighed as he chose to give up the security and privilege of his position and independently plot the overthrow of a system from which he benefitted. Perhaps Charles's mother whispered to him the story of her own rape, or inculcated in him a sense of rage and resentment toward the white planter class. Perhaps the sons and brothers of the Trépagnier family had Charles's woman for sport. Perhaps Charles could no longer consent to savagely beating his fellow slaves. Perhaps he could not bear the resentment, jealousy, and bitterness of all those who labored eighteen hours a day in the field under his command and management.

Whatever motivated him, Charles kept his rage and his plot as secret as possible. Had even the slightest hint of his plans for betrayal leaked out, Charles would have faced instant execution—such was the price of insurrection. He had to lie to both his white master and his slave subordinates on the plantation, letting both groups think that he was a contented and successful driver.

* * *

As the white planters celebrated Epiphany and prepared for the night's celebrations in New Orleans, the planter James Brown took mental note of a meeting between three slaves from three of the wealthiest plantations on the German Coast. Thinking that no one had noticed their absence amid the festivals, the three men gathered on the plantation of Manuel Andry, forty-one miles northwest of New Orleans. Cramped into the small space of a dilapidated shack behind the mansion house, the three men talked in hushed voices. Charles Deslondes gazed nervously out the window. His eyes looked to the second-story piazza of the large Spanish Colonial mansion as he instinctively checked for the presence of Andry. These two coconspirators were some of the only ones he could trust to know his secret purpose.

Now he listened intently to the African rhythms of Quamana's speech. Quamana's face bore the blood markings of the Akan—he was an intimidating man. Captured and brought across the Atlantic Ocean a mere five years before, he was Kook's best friend and close associate. Sick of the brutal work of sugar planting, Quamana perhaps now talked of what one slave would later describe as the goal of the uprising: to kill all the whites. Charles had heard this talk before, having met with Kook and Quamana frequently on his trips to the Trépagnier estate.

Twenty-six-year-old Quamana was a slave at James Brown's plantation, located ten plantations downriver from the Andry estate. His political radicalism and dedication to the cause must have proved inspirational to many other slaves, his enthusiasm as contagious as the deep anger from which it emerged. He had traveled from plantation to plantation through the dark cypress swamps on the edges of the cane fields. Posting a spy in a tall tree to watch for intruders, just as some of them had done as children in Africa, the Akan warriors in Louisiana met to organize the uprising.

The third slave at the meeting was Harry Kenner. Originally from Virginia, he had developed a trusted core of English-speaking slaves—a dozen slaves on his plantation who would participate in the revolt. A twenty-five-year-old carpenter, Harry was a slave at the Kenner and Henderson plantation, twenty-one miles to the southeast at the end of the German Coast closest to New Orleans.

These three men, each with different insights and abilities, had planned their insurrection and spread word of the uprising through small cells distributed up and down the coast, especially at James Brown's plantation, the Meuillion plantation, and the Kenner and Henderson plantation.

These cells were born out of networks of communication that tied the slaves to New Orleans and its diverse marketplace and ports. During their free time on the weekends, slaves often participated in the thriving economy of the region. They grew staple crops, raised small livestock, and collected wood and moss, and traded these products of their labors to itinerant peddlers or in the marketplaces. Black peddlers went door to door marketing goods. The River Road was full of activity, whether organized by the masters or the slaves, and these activities formed the base of the revolutionary cells.

What did these men talk about in their secret meetings, behind the closed doors of the slave cabins or under the tall trees on the edges of the fields? They wrote nothing down and told no one. But all evidence points to a revolutionary ferment. The slaves, it seems, were growing increasingly radical in their political views—a radicalism that occasionally bubbled up into outright violence.

Prior to the sugar boom, New Orleans was a poor, multicultural city with very few social controls. The lines between slavery and freedom were not clearly drawn, and slaves frequently escaped into the swamps to form maroon colonies. There was a history of armed resistance in these areas that drew on French, Creole, and Kongolese traditions. These insurrectionary traditions shaped the lives of the slaves and represented an alternative political culture to that of the planters.

In the 1780s, the slave Juan Malo from the d'Arensbourg plantation on the German Coast led a thriving maroon colony in the swamps below New Orleans. St. Malo, as he named himself, was reported to have buried his axe into a tree near his colony and declared, "Woe to the white who would pass this boundary." St. Malo and his men—reportedly numbering over 100—repeatedly repelled the raiders sent by the Spanish government who came into the swamps on pirogues armed to the teeth with guns. The maroons built extensive networks of slaves on the plantations that provided them with food and tipped them off about impending raids. Eventually, the Spanish grew so incensed by St. Malo's independence and the threat he posed to the slave plantations that they sent a massive force of militiamen into the swamps in 1783. The militia, following the tip from a spy, came upon the unsuspecting maroons and opened fire. This time their expedition succeeded. They captured a wounded St. Malo and brought him back to New Orleans. On June 19, 1784, the Spanish hanged St. Malo in the center of New Orleans—creating a martyr and a folk hero for the German Coast slaves.

In 1795, the Spanish discovered a massive slave conspiracy at Pointe Coupée—an area on the high grounds between New Orleans and Natchez. The conspiracy took place at the height of the French Revolution and just after the slaves in Saint Domingue had forced France to abolish slavery. The planters discovered the book *Théorie de l'impôt*, including the Declaration of the Rights of Man, in the cabin of one of the slaves (the Declaration, adopted by the French national assembly in 1789, declared that all men are born free and equal and maintained the rights of "liberty, property, security, and resistance to oppression"). Several slaves reported hearing rumors that the slaves had been freed in the colonies—one even specifically mentioned Saint Domingue. The slaves planned their uprising at church during the Easter holidays and Holy Week. They also held meetings at the slave quarters of different plantations and in the marketplaces. The plot was discovered, however, before it ever came to fruition. The planters hanged twenty-three slaves, decapitated them, and nailed their heads to posts. They flogged thirty-one additional slaves and sent them to hard labor at Spanish outposts in Mexico, Florida, Puerto Rico, and Cuba. By the time of the Pointe Coupée uprising, the revolutionary fervor of the age had reached the River Road, inspiring the slaves to Jacobinism and an assertion of their rights to freedom.

In 1805, after two years of American control, there were rumors of another slave conspiracy. The residents of New Orleans were alarmed to discover a traveling Frenchman preaching the revolutionary philosophy of liberty, equality, and fraternity to the French-speaking slaves of Louisiana. The government promptly arrested this dangerous man, bringing great relief to the planters. The planters knew they had much to fear from such loose talk of revolution.

While French and Creole maroon influences were strong in the Orleans Territory, there was also a huge influx of Africans to the area. Slave traders brought around 20,000 African slaves to New Orleans between 1790 and 1810. These immigrants brought with them their own violent history. Kongo, the source of over 10 percent of these slaves, was going through revolutionary contortions just as France and Britain were. Kongo had been ripped apart by civil wars, producing thousands of veterans trained in military practice and willing to use force to obtain political ends. Many of these well-trained war veterans were sold off into slavery and spread throughout the New World, disseminating their warrior knowledge. The Kongolese had developed their own style of warfare, a form of guerilla tactics, that involved spreading out over space, quickly retreating in the face of threats, and using ambushes and terrain advantages to the best of their abilities. They used flags and drums to rally the troops and communicate in the field of battle.

Charles and the other slaves on the German Coast were well armed with revolutionary ideology, and some with military training. They were conversant in the doctrines of the French Revolution, and aware of the powerful example of the Haitian revolutionaries. They drew on significant Kongolese and Akan populations trained in guerilla warfare and experienced in the use of violence for political ends. The sporadic rebellions of the last few years were like the beat of a drum slowly building to crescendo. And it seemed that the only ones who hadn't heard the music were the planters. Though terrified by Haiti, the planters refused to acknowledge or try to understand the political logic behind the slaves' actions. Their racial ideology and pride in their own accomplishments led them to miss all the warning signals of the impending revolt.

* * *

As Charles, Quamana, and Harry met on that Sunday in 1811, they came well armed for battle with a powerful set of revolutionary political ideas, well-honed skills, and a complex organization of insurrectionary cells prepared to attack as soon as they gave the word. And by all accounts, they gave the word that day. Thursday would be their moment to strike. And on that Thursday, this diverse group of slaves would mount the greatest challenge to planter sovereignty in the history of North America. On January 8, 1811, they would turn their world upside down.

January 8, 1811

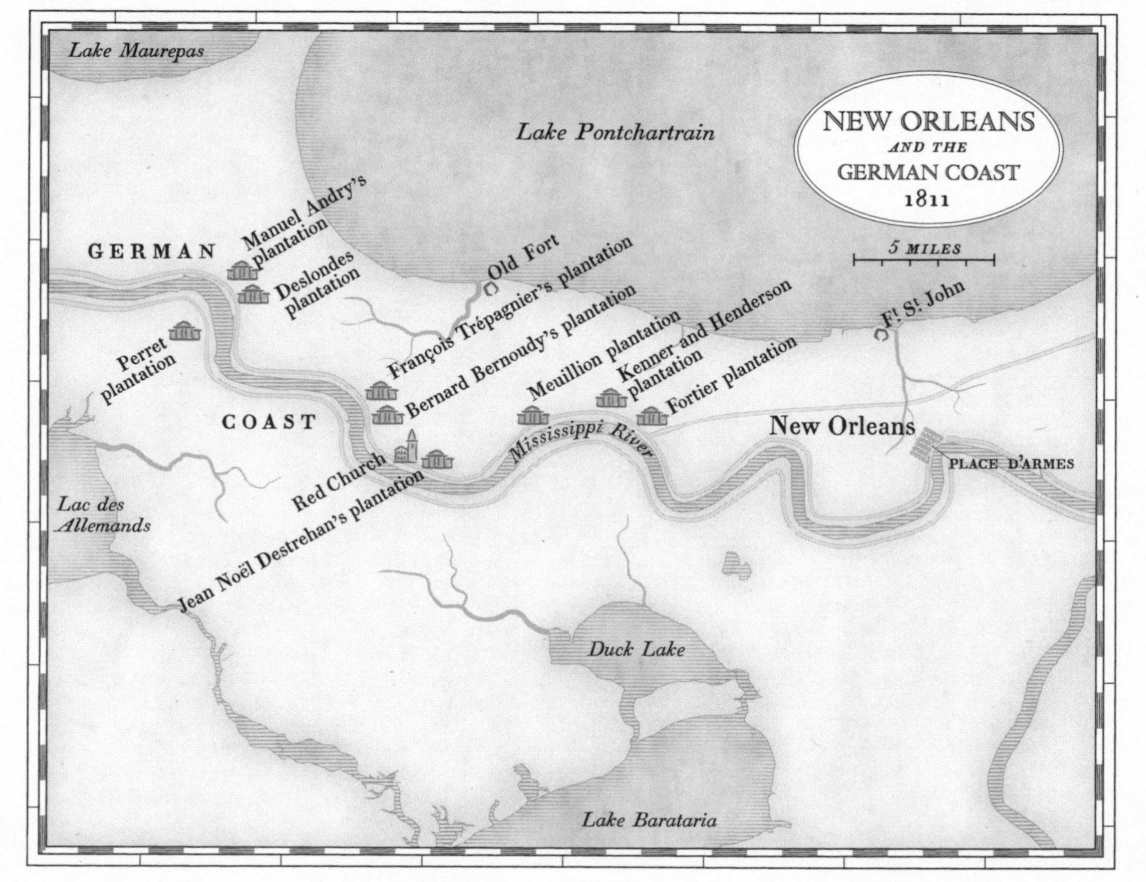
NEW ORLEANS
AND THE
GERMAN COAST
1811
5 MILES
Lake Maurepas
Lake Pontchartrain
GERMAN
COAST
Manuel Andry's plantation
Deslondes plantation
Perret plantation
Old Fort
François Trépagnier's plantation
Bernard Bernoudy's plantation
Meuillion plantation
Kenner and Henderson plantation
Fortier plantation
Red Church
Jean Noël Destrehan's plantation
Mississippi River
Ft. St. John
New Orleans
PLACE D'ARMES
Lac des Allemands
Duck Lake
Lake Barataria

Eight
REVOLT

For the planters and the slaves alike, January was a time of celebration. The sixteen-hour workdays of grinding season were over, and lavish parties in the homes of the planters and in New Orleans marked the celebration of Christmas and Epiphany. January also marked the onset of Louisiana's winter. Some time in the first few days of the month, storms from the northwest blew in a powerful rainstorm. By January 6, the roads were "half leg deep in Mud." Rain meant even more time off work, because excessive rain flooded the soil, making movement difficult and making it nearly impossible to work the soil or haul wood from the swamps. The slaves, then, were idle—the most dangerous state, from the perspective of the slave owner.

On the night of January 8, the rain continued to come down. Water coursed along the wood roofs of the slave quarters, drowning their staccato voices with streaming, rushing noise. Twenty-five dark faces looked on as the slave driver turned rebel Charles Deslondes laid out the plan and gave some final words of encouragement. Every man assembled knew that his presence meant a near-certain death sentence if the revolt failed. No slave revolt in Louisiana had ever before been successful, and the punishment for failed rebellion was clear: torture, decapitation, and one's head upon a pike. Yet with the planters distracted by Carnival and the American military fighting the Spanish past Baton Rouge, the slaves believed they just might have a chance.

No records survive to tell us what Charles said to his men in the final minutes before they attacked. The slaves were preparing for battle, not taking notes. But perhaps Charles acted like the leader of the 1812 Aponte Rebellion in Cuba, who took a sharpened machete, stabbed it through a plantain, and shouted to an audience of slaves, "This is how I will run it through the stomachs of the whites."

As the slaves made their final preparations, their planter masters, Manuel Andry and his son, Gilbert, lay asleep in their beds in their respective chambers, surrounded by family portraits and fine furniture from Europe. Lulled to sleep by the pitter-patter of rain, perhaps they now dreamed of the month of dances and parties ahead. They felt secure in a world they had created. Both leaders of the colonial militia, the two were respected men in their community. But to the slaves, they were known only for their cruelty—for the frequent whippings that left deep scars in the backs of several of the newly minted rebels and the iron collars they would fasten around the slaves' necks.

With the clouds darkening the cane fields and the rain blotting out the noise of their approach, the slaves hastened toward the back door of the Andry mansion. Catching each other's eyes glinting in the night, they held their cane knives and machetes with tight fists. Even in the darkness, Manuel Andry's plantation cast a formidable shadow. A high roof soared into the sky, shielding a piazza and a broad gallery from the rain. With Charles leading the way, the slaves entered the brick-walled storage basement and made their way toward the wooden double staircase that led upstairs to the quarters where Manuel and Gilbert Andry slept.

As the slaves stormed onto the second-floor landing, Manuel Andry woke to the sight of dark forms penetrating his bedroom. As his eyes snapped open and his brain awoke with a fright, Andry caught a glimpse of Charles Deslondes, a new look on his face, ordering his fellow slaves toward Andry with an axe. One can only imagine Andry's reaction, in the fog and panic of those first instants of awareness, to seeing Charles, his most loyal driver, his reliable assistant for over a decade, the man he had trusted to manage his plantation, now turned betrayer and potential murderer.

His mind clouded by fear and anger, Andry's eyes fixed on Charles's axe, a plantation tool transmuted into an icon of violent insurrection. As the slaves surged toward him, Andry leapt from his bed. The slaves stood between him and the staircase—and the staircase was his only way of escape. Andry made the decision to act, charging toward the surprised slaves.

As he rushed through the crowd of rebels, the slaves lunged at him, slicing his passing body with three long cuts. But somehow Andry made it past. He hurled himself toward the staircase, turning his head only to catch a most horrifying sight: the slaves swinging their axes into his dying son's body.

Pursued by a pack of angry rebels, Andry could not stop. He could not turn back. With the bloodcurdling vision of his son's death emblazoned in his mind, his adrenaline took over. He ran for his life. He sprinted through the clover fields in front of his mansion toward the water, where he knew a pirogue lay on the levee.

As the slaves hacked Gilbert Andry into pieces, Charles decided that it would be fruitless to send men chasing after Manuel. His ambitions were greater than killing one planter—even a planter he hated so personally. He sought liberation and conquest on a greater scale. He did not think Manuel Andry would make it too far—and even if he did, a wounded middle-aged planter posed little threat to his slave army. Or so Charles thought.

In Charles's mind, the tide had finally turned. Baptized with the blood of his former master, Charles and his men broke into the stores in the basement of Andry's mansion, taking muskets and militia uniforms, stockpiled in case of domestic insurrection. Many of the slaves had learned to shoot muskets in African civil wars, while others would fight more effectively with the cane knives and axes they had learned to wield in the hot Louisiana sun. As his men gathered weapons and shoved ammunition into bags, Charles and several of his fellow slaves cast off the distinctive cheap cotton slave clothes and put on Andry's militia uniforms. Charles knew that the uniforms would lend the revolt authority, wedding their struggle with the imagery of the Haitian revolution, whose leaders had famously adopted European military garb. As they sought to rally other men to their cause, he must have hoped the uniforms would reassure the doubters of the legitimacy of their plan and their organization. If the revolt were to succeed, he would need numbers.

Amid the rainstorm, Charles shouted orders to his fellow slaves. They assembled in the clover field in front of Andry's plantation, falling into line behind Charles, who was now mounted on horseback. They were familiar with military discipline: their work on Andry's sugar plantation had taught them to follow orders with alacrity. But now they were motivated not by fear of the lash, but by the hope of freedom. They were forty-one miles from the gates of New Orleans, which they hoped to conquer in two days' time. Asked later why he had left the Andry plantation that night, the rebel Jupiter replied that he wanted to go to the city to kill whites.

Charles and his men began to march. Charles shouted, "On to New Orleans!" and the newly formed rebel army shouted it right back. The revolt had begun. As the twenty-five rebels gathered guns, knives, and horses on the Andry plantation, rumors of the insurrection's inception flashed like lightning through the German Coast. In those dark early-morning moments, the slave quarters for miles around erupted. Slaves ran from door to door, whispering the news, and small conferences gathered in tight quarters as men and women weighed their options: to risk death and join Charles and his men or stay behind in safety.

Charles and his fellow leaders had planted the seeds for the revolt well over the past few months. In cautious conversations on the edges of the sugar fields, in the Spanish taverns along the levees, and at the weekly Sunday dances, Charles had built a strong organization. Inspired by the stories of the Haitian revolution and flush with the philosophies of the French Revolution, the diverse band of slaves that joined insurrectionary cells believed they could secure freedom, equality, and independence through violent rebellion. As the heads of the whites rolled through the streets, they could form a new republic—a black outpost on the Mississippi, guaranteed by force.

Armed with plantation tools and primed by revolutionary ideals, roughly one-quarter of the slaves on the plantations along the River Road gathered on the levee to meet the marching rebels and join the insurrection. In those predawn hours, the slaves shivered with cold and anticipation, the rain soaking their cotton clothes. In military formation, the slaves marched along completely flat land on a well-trodden road toward New Orleans.

To the right, the rough waters of the Mississippi surged by past the four-foot-high levees. To their left stood the plantation mansions. Live oaks, trimmed to regular shapes, young orange trees, deciduous Pride of China trees with bursting yellow flowers, and other tropical trees and bushes decorated the plantation lawns. Side roads marked by avenues of laurel trees cut into pyramids led to the mansion houses that adorned each plantation. Behind the mostly two-story-high mansions, with their piazzas and covered galleries, stood the slave cabins—and behind the slave cabins, the sugar fields stretched for a mile before finally giving way to the marshy cypress swamps.

In front of them lay the road they would take on the two-day journey to the city gates. Wooden bridges covered the worst spots, the ditches that provided irrigation for the fields. But in many places the road was almost knee-deep in mud. As the road wound its way toward New Orleans, it passed dozens of plantations in quick succession.

The rainstorm could not drown out the feelings of pride and power the rebels felt as they looked in front and back of them and saw the ranks of the committed swell. The beating of an African drum keyed the men to excitement. About five miles down the road, as they rounded another turn of the Mississippi River, the rebel army saw a heartening sight through the fields where only a month before sugar cane had stood higher than a man's head. A group of ten slaves stood under the tall oaks fronting the plantation of the local judge, Achille Trouard. The slave Mathurin now sat on one of Trouard's horses, commanding a group of about ten slaves. Waving his saber in the air, Mathurin pledged his troops to Charles's cause. The two leaders formed the heart of the incipient cavalry—leaders on horseback that would snowball into a full-blown troop as the men proceeded closer to New Orleans.

Over the sound of the horses' hoofs, Mathurin shouted some bad news. Achille Trouard, his master, had heard about the revolt before he and his men could attack. Led by a loyal house slave, Trouard and his two nieces had fled into the swamps to hide. The rebels had made their first moves, but they were not the only players.

* * *

The slave Pierre awoke Alexandre Labranche around dawn, not to kill him, but to save his life. Another slave, François, rushed in a few moments later, advising Labranche to "flee immediately into the woods back of [the] farm."

François had heard about the revolt just as many newly minted insurgents had—through the grapevine telegraph of news and information that coursed through the slave quarters. But François had made a different calculus from his more hot-headed comrades. He chose to betray the revolt, most likely not for any love of the planters or of being a slave, but out of pragmatism. All the odds were against the slave-rebels, and François knew well that his best chance of survival amid the brutal work of sugar planting came from siding with the powerful and pledging allegiance to the white men who controlled the entire world as he knew it, from the slave ships that picked him up in Africa to the great city of New Orleans to the small worlds of the plantations.

François was not alone. The majority of slaves chose not to fight. All knew that the clearest path to freedom was not to join a revolution but to betray one; the planters had made that much clear. In the face of such might, most slaves decided that rebelling was simply not worth the cause, and some sought to profit by betraying the rebels they believed were making a foolish and wrongheaded decision. So François chose to tell his master, Alexandre Labranche—a decision that might even mean the reward of freedom if Labranche survived.

In a panic, Labranche awoke his wife and children and

gathered a few key possessions—a musket, some warm clothes, and a few biscuits. Then they hurried downstairs to meet François. Casting backward glances to the levee and the River Road, François and the Labranches hurried through the fields toward the cypress swamps, and the dark trees that would hide them until the revolt had passed. The planter family and their black guide fled in terror along paths first blazed by escaping slaves.

Other planters did the same. In what one planter described as a "torrent of rain and the frigid cold," the white elite of the German Coast flew on horseback to New Orleans, concealed themselves in the swamps, or took boats to the other side of the river where order still prevailed. Hermogène Labranche (Alexandre's brother) and his family holed up in the woods until the slave-rebels passed. They then took a boat to the other side of the river. Adelard Fortier escaped to New Orleans. As they left their plantations, the planters handed over authority to black slaves they trusted to be loyal—ordering these men to guard their mansions and prevent the slave army from wreaking destruction on their stockpiles of sugar.

* * *

The planters' decisions to flee emboldened the rebels. The slaves were not used to seeing black men using violence to control white men—and the effect was intoxicating. "Freedom or death," they shouted. Impressed slaves rallied to the rebel flags.

Nowhere was this new-felt power more evident than on the plantation of James Brown, only a few miles southeast along the river from the Andry plantation. Here the Akan warriors Kook and Quamana watched as their master mounted his horse and spurred it along the road toward New Orleans. As his figure vanished into the rain and mist, Kook and Quamana must have spoken of a government of black men, the execution of the white planter class, and the impact their weapons would have on the Mississippi River colony of New Orleans. This was not the first time these slaves had heard such talk, but Brown's departure reassured the wavering that this might be their great opportunity to finally realize what only days before had seemed to be mere wild dreams.

Inspired by Kook's and Quamana's silver tongues and fiery passion, half of James Brown's slaves chose to join the seventy-five rebels who arrived soon after their master's departure. Kook, Quamana, and their men formed a new, more radical core to the slave insurgency. At over six feet tall, with the etchings of an Akan warrior inscribed in his face, Kook towered above his fellow slaves. At the time, the average height of African males was 5 feet 3 1/2 inches, so Kook must have stood out as a giant. His rippling biceps, swollen from years of swinging a machete in the cane fields, made him a force to be reckoned with. He commanded easy and instant respect, even among the most hardened and rebellious slaves.

The slaves on the Brown plantation would have known Kook and Quamana both for their Asante heritage and their political fervor. In 1806, the same year Kook and Quamana were sold into slavery, the Danish governor of the region wrote, "The Assianthes, with sword and fire murder and destroy everything they meet with . . . A great battle was lost in the ninth of this month, at which four towns were laid in ashes and many thousands of people, not shot in battle, but murdered in cold blood." Perhaps Kook and Quamana told their fellow slaves of these great battles, assuring them that New Orleans would soon burn like the defeated towns of subdued African tribes.

With the addition of Kook and Quamana's detachment of Akan warriors, the slave army now numbered well over a hundred men. Beating drums and shouting with the joy of freedom, they urged each other on. They were a motley crew. Drawn in small groups from over a dozen plantations, the slave army boasted men from Virginia, Kentucky, Louisiana, Cuba, Senegambia, the Asante kingdom, and Kongo. Eleven separate leaders rode on horseback, each representing a different ethnic faction: the Muslim Senegambians, the Akans, the Sierra Leonians, the French Louisianans, and the Anglo-Americans were all involved.

With the evacuation of most white planters, the slaves had met few serious tests of their resolve. Charles and his fellow leaders must have worried that when faced with armed white men, their army might melt away. They must have hoped the uniforms, the disciplined regiments, the flags, and the drums would add confidence, but they were in uncharted waters.

The slave army had another problem as well—continuing betrayals. The slave Dominique, who belonged to Bernard Bernoudy, was staying at the Trépagnier estate with his wife. Hearing of the revolt from slaves who lived near his wife, he saw an opportunity to enhance his own status among the planters. Dominique rushed to tell François Trépagnier that "there was a large number of rebel slaves moving down the river, pillaging the farms and killing whites." After warning Trépagnier, Dominique departed, ostensibly to warn Bernard Bernoudy of the impending danger. On his way to Bernoudy's plantation, Dominique passed through the plantations of Delhommes Rilleaux, James Brown, Pierre Pain, and Alexandre Labranche, where he passed the word directly or through enslaved intermediaries. When Dominique arrived back at Bernard Bernoudy's plantation, Bernoudy sent Dominique to New Orleans, who alerted whites along the way. Most would choose to flee.

But despite Dominique's warning, François Trépagnier made the decision to stand his ground in the face of men he considered little better than canines. His contempt for his slaves was well known. Local legend had it that he kept a slave boy named Gustave as a house pet. As Trépagnier ate, he would toss food from the table onto the floor, where Gustave would pick it up and eat it. Other men had dogs, went the story, but Trépagnier had a black child.

Trépagnier did not think the slave-rebels would pose much of a threat. His wife and children begged him to accompany them into the swamps to hide, but he refused. Sending them off, he resolutely loaded his fowling pieces with buckshot. Amid the bitter chill of January rain, he locked his doors and took up a position on the second-floor gallery. There he waited for the slaves to arrive.

Trépagnier did not have to wait long. In the morning light,

he could see the smoke from four or five burning plantations curling into the sky. He could hear the ominous beat of the African drums, pounding and pounding. But Trépagnier was not afraid. He had heard similar drums at slave dances, and he imagined he would be more than a match for a ragtag band of slaves. Quickened by the shouts and calls of the rebels, Trépagnier did not expect to see what he saw next.

Around the bend of the levee came a veritable army. Divided into companies, each under an officer, black men in militia uniforms advanced toward his plantation. Marching along the levee, they shook their fists and their weapons in the air. At the sight of the slaves, Trépagnier leveled his double-barreled gun and began to fire. The smoke of his weapon engulfed the second-floor gallery that skirted his mansion. Buckshot was notoriously inaccurate, and the smoke obscured the shooter's visibility. But Trépagnier hoped the slaves would be intimidated by his presence and turn back.

Trépagnier's estimation of the slaves' strength and his decision to stay behind proved dead wrong. Kook led a party behind the plantation house, up to the second story. As other slaves lit fire to the foundations, Kook took his axe and hacked François Trépagnier to pieces. Local legend has it that Gustave too swung an axe, exacting final vengeance for years of patronizing mistreatment.

The rebel army had passed its first test. They had annihilated one of the most hated men on the German Coast. But Charles and Kook and all the others knew they needed more blood to baptize their incipient revolution. The slave army knew that the obstacles remained extreme. The leaders of the revolt knew the stories of prior acts of resistance—and their consequences. They knew what was at stake. No white man, no American official, no French planter, would brook the survival of a black army anywhere near white power centers. Slavery was too intertwined with the political economy of the Atlantic world to allow for any sort of black political existence.

Only through extermination and extreme violence could they earn the right to form a separate political community—to be recognized as men rather than slaves. Violence was simply the price they had to pay for freedom, and Gilbert Andry and François Trépagnier were the down payment.

Charles, Kook, Quamana, and their compatriots hoped that their swelling army would grow exponentially as they neared the dense and rich plantation zone right around New Orleans, and their aim was to conquer the city from the weak American military force there, forming a base to which slaves from far and near could flock. Inspired by the Haitians' victory over Napoleon's army, the rebel leaders were not unreasonable in imagining that they could defeat the Americans' small colonial army.

They had driven their masters into hiding and thrown down a bloody gauntlet to the authority of the sugar masters, burning down several sugar plantations. Now, with an army several hundred strong and buoyed by victory, the slaves continued their march toward New Orleans. As the sun began to light up the fields from behind the storm clouds, a fearsome rebel army stood just outside the most densely populated plantation region in North America. In just a few miles, the slaves would arrive at the Red Church, then the Destrehan plantation, and then the new American plantations erected just outside of the city, with their huge populations of recently purchased Africans. Fear, rage, and violation lay as deep as the January mud, and the clash between the incipient forces loomed imminent on the horizon.

January 9, 1811

Nine

A CITY IN CHAOS

Morning came slowly in the city. In a driving rain, the sun could only slowly illuminate the dirt streets and brick sidewalks of New Orleans on the morning of January 9. The white spires of the cathedral and tall masts of the ships crowding the harbor topped the center of the city. In the dense neighborhood around the Place d'Armes, small brick houses two or three stories high clustered about grand old Spanish houses. Once a palisade and a ditch ran around the center of the city, forming a parallelogram with the river. Four redoubts stood at the corners to protect the city's inhabitants—though all but the fort at the entrance of Faubourg Marigny had since been demolished. Since the American acquisition, the ditch had been filled up and planted with trees, leaving a ring of open space between the city and the suburbs. A boulevard called Rue de Rampart ran where the ancient town wall used to stand. Parallel to the river, roads lined with reflecting lamps passed from the center of the city out toward the plantation zone to the northeast. Here the old houses of the present-day Garden District gave way slowly and almost indistinguishably to the rich sugar plantations of the German Coast.

The black slaves of New Orleans were the first to arise on January 9. As their masters slept, the slaves began preparing for the day: readying the horses, lighting the fires, and cleaning the houses. Slaves formed the great underclass of the city, serving as laundrywomen, delivery boys, cleaners, construction workers, carriage drivers, cooks, peddlers, stevedores, boatmen, and manual laborers of all sorts. Slaves walked hurriedly along the city's brick sidewalks. The rain had washed out the roads, leaving prodigious quantities of mud and large puddles where hard-packed dirt usually lay. In crossing these watery streets, the slaves faced a difficult task. They could either search for stones properly placed for jumping or risk sticking fast in the mud as they waded from one side to the other. Few of these early risers knew just what the day ahead would hold.

* * *

The military scouts at the western gates of the city were the first to hear the news. The warning came from galloping horsemen who notified the garrison of the revolt. But they could not say how close the slaves might be or just what was unfolding. Within hours, fleeing refugees in carts and carriages crowded the road into the city, forming a traffic jam nine miles long.

Recalling the scene a few days later on January 17, a correspondent to the Louisiana *Gazette* described a road that "for two or three leagues was crowded with carriage and carts full of people, making their escape from the ravages of the banditti—negroes, half naked, up to their knees in mud with large packages on their heads driving along toward the city." Rumors ran rampant through the city. "The accounts we received were various. Fear and panic had seized those making their escape and it was not possible to estimate the force of the brigands."

Fresh memories of Haiti fueled panic and terror among the city's white inhabitants. They had heard the stories of the Haitian revolution, when rebel slaves strapped white planters to racks and cut them in pieces, raped their daughters and wives, and decapitated men, women, and children alike. They feared that the German Coast had become, as one resident put it, a "miniature representation of the horrors of St. Domingo." Women and children fled through the streets toward the redoubt at Faubourg Marigny to take shelter with the small garrison located there.

Sometime before noon, Governor Claiborne heard the news from his top general Wade Hampton, who had arrived a mere two days earlier to help with an ongoing war with the Spanish over West Florida. His first thought was for the safety of the city. He feared not only that the rebel army would arrive before his troops could prepare, but also that there might already exist some communications between the rebels and the urban slaves and free black people. In a city where the majority of the population was black, he feared the opening of a second front. If the rapidly closing rebel army were able to take advantage of an urban slave riot, Claiborne knew white New Orleans would stand little chance of survival.

Holed up with the mayor and other officials in the government buildings surrounding the Place d'Armes, Claiborne dispatched orders to the militia and the military to seal the city. His first terse writings on January 9 were to General Hampton: "Sir, I pray you to have the goodness to order, a Guard to the Bayou Bridge, with instructions to the Officer to permit no Negroes to pass or repass the same." He wanted to prevent the flow of information from the black residents of the German Coast to the black residents of New Orleans, to quarantine the city from the contagion of revolt.

After securing the bridges, Claiborne targeted the taverns next. "All the Cabarets in the City and Suburbs of New Orleans are ordered to be immediately closed," he decreed from his headquarters. Claiborne feared such halls of entertainment provided sites not just for the mixing of the lower classes, but also for the spread of revolutionary ideas. He was not the first to make this connection. A 1781 Spanish report detailed the troubles caused by black men during the Carnival season. "People of color, both free and slaves, were taking advantage of carnival," wrote the Spanish official, to go about "disguised, mingling with the carnival throngs in the streets, seeking entrance to the masquerade balls, both society balls and those charging admission, and threatening the public peace by introducing enemies of the king into assemblies under mask," and even committing robberies.

Claiborne put the city into lockdown. "No male Negro is permitted to pass the streets after 6 o'clock," he ordered. The city garrison would fire a gun at dusk—the final warning to any black man still in the streets. The gun shot left little to the imagination of what would happen to any male slave found outdoors at night.

With these orders in place, Claiborne set his sights on the slave army now approaching his city. As reports of atrocities spread through the city, Claiborne turned to General Hampton. As Hampton later recalled, "about 12 O'Clock on the Morning of the 9th the governor came to me with the unpleasant information that a formidable insurrection had commenced among the blacks, on the left bank of the river, about 40 Miles above this city, which was rapidly advancing toward it, and carrying in it's train fire, Murder, & pillage. The regular force in the City was inconsiderable, and as there was nothing like an organized Militia, the confusion was great beyond description." Over the next six hours, Hampton scrambled to respond to Claiborne's request for help. By six o'clock he had marshaled two companies of volunteer militia and thirty regular troops to meet the rebels. With the exception of a small garrison at the fort, this small force was the extent of American military power in New Orleans. The company set out after sunset along the River Road to face what some of them feared was a slave army of equal ferocity to the revolutionaries of Haiti.

Commodore John Shaw, the naval commander of the fleet at New Orleans, expressed skepticism of General Hampton's force, calling it a "weak detachment." Shaw feared that the insurgents might triumph over Hampton's troops, that "the whole coast [would exhibit] a general sense of devastation; every description of property [would be consumed]; and the country laid waste by the Rioters." The vulnerability of New

Orleans contributed to the sense of panic. The majority of the American military force (in particular the highly effective dragoons) were in West Florida. With the departure of the soldiers, the volunteer militia, and the seamen, New Orleans was left virtually defenseless. "All were on the alert . . . General confusion and dismay . . . prevailed throughout the city," Shaw wrote. "Scarcely a single person in it possessed a musket for the protection of himself and property." The slave-rebels had forced the utter evacuation of military power from New Orleans. And now they faced the sum total of the military might of the Orleans Territory—at this time a mere sixty-eight regular troops.

Driving wind and a steady rain prevented armed ships from moving up the river. Bur Commodore Shaw made the quick decision to arm the sailors and send them in support. He lost no time in attacking by land. He sent his lieutenants Charles Thompson and Harvey Carter to lead a detachment of forty seamen on the expedition. The combined force now reckoned about 100 men—barely 20 percent of the size of the slave army. Moreover, they had little sense of the terrain, having come recently from the East Coast. This small detachment seemed to have little chance of success.

With the fate of the city in the hands of the army and the navy, Claiborne sat down as night was falling to draft his official reports. Claiborne knew he was in danger of losing control over the city he had governed since 1804. But the matter was out of his hands; he would have to rely on Hampton and Shaw to defend the city. Religion was his last resort. "I pray God that the force sent from this City may soon meet the Brigands and arrest them in their murdering career," he wrote late on the night of January 9. With the white residents of the area clustered behind the city gates and the black slaves marching from the fields, it seemed all Claiborne could do was pray.

* * *

While Claiborne took frantic action to secure the city, his soldiers began to encounter frightened fugitive planters. But to the soldiers' surprise, some planters chose to fight rather then flee. As they met the American army, they turned their horses around, facing back now toward their homes and the slave-rebels. Though lacking leadership, the planters formed a party of volunteer cavalry and agreed to join in the attack. Not waiting for Hampton's troops, the horsemen led the way along the River Road and into what was now enemy territory. Many of these men were from families who had built New Orleans; they were patriots and lovers of their homeland. And now, in the face of perhaps the city's greatest challenge, they rode out to defend it. In their minds, the slave-rebels were not freedom fighters but terrorists. As they passed scores of refugees heading toward safety through the pouring rain, they soon had to answer the questions posed by the revolt—and the changed world, post–Gilbert Andry and François Trépagnier's deaths.

Riding along the River Road, the planters heard the clip-clop of a lone horse heading toward them at a fast clip. Suspecting the rider might be another escaping planter, Jean Noël Destrehan, Alexandre Labranche, and René Trudeau rode out in front of the troops to greet and debrief with the comrade they expected to emerge. But as the rider approached closer, they realized the horseman was black. Taking out their guns, they ordered the black rider to slow his horse. With no other clear option as he rode into an army of white planters, the slave brought his horse up next to the planters. He was unarmed. René Trudeau, recognizing his own slave Jacob, "stopped near the said negro, and said jokingly: 'As brigand are you not at the head of the negroes?'" It was a loaded question—Jacob's answer would make the difference between life and death. Jacob sought to defuse the charge. "My master, you know me, that I am not capable of such a thing," he responded. Jacob's choice of phrases was deft, born of long experience with the charged pleasantries and lies of plantation life. He had emphasized his state of servility by addressing Trudeau as "my master." Jacob next asserted his familiarity, saying, "you know me," relying on a mutual acknowledgment of good intentions, on a reputation for loyalty that he had presumably built up with Trudeau. Finally, Jacob denied that he would ever participate in a revolt. "I am not capable of such a thing," he told Trudeau. What he implied, however, was not merely that he could not possibly fathom revolting, but that he lacked the facility or agency to do this. Having heard Jacob's response, Trudeau decided to spare his life, to merely imprison Jacob until the revolt was over and a trial could be held to determine what exactly Jacob was doing. With Jacob in chains, the planters continued their ride toward the chaos of the German Coast. Their encounter with Jacob left them again feeling in control—as though the rebels they would soon face were just the same slaves they had known for years.

Ten

A SECOND WIND

As the day waned, the rebels were confronted with a new reality. They found each plantation home they came upon empty except for the slaves. The planters they had intended to surprise and kill were gone. As the chaos of insurrection had spread along the German Coast, the balance of power had shifted into the hands of the slaves. The planters no longer felt safe in their homes, in the flat, visible space between the river and the swamps. But as much as Charles, Kook, Quamana, and the rebel army reveled in the speed and efficiency of their conquest, they knew they had not heard the last of the planters. They also knew they had to strike hard and fast in order to achieve their goals. They would need an early victory against a substantial planter force in order to persuade wavering slaves to join them and to ensure ultimate victory.

The new recruits were bursting with energy, but the long walk was taking its toll on others. The slaves who had joined the rebellion at Manuel Andry's estate were feeling the long march in their legs. But fortunately for these tired souls, Bernard Bernoudy kept a substantial collection of horses on his estate. As they marched into the plantation, the slave Augustin, a highly valued sugar worker, drove the horses toward the slave army. Horses were powerful military tools, enhancing the speed, power, and stature of the slaves. With the infusion of these new beasts of burden, about half the slaves were able to ride instead of walk, accelerating the pace of the rebels' progress toward New Orleans and increasing their standing in the minds of the slaves they met as they proceeded further.

At the plantation of Butler and McCutcheon, the slave Simon was waiting. Simon had grown up with his family in Baltimore. But when Simon was in his late teens, his old master had sold him down south to New Orleans—forcing him to leave behind a family he would likely never see again. The twenty-year-old slave had tried to escape just months before, to flee and rejoin his family in Baltimore. But Butler and McCutcheon were well-connected slave masters, and after they placed an ad in the local newspaper, Simon was quickly apprehended and returned to the German Coast. There he was savagely beaten for the transgression of attempting to reunite with his family. Scars on his left cheek and his forehead marred his handsome features. Simon had rallied eight other young men in their twenties—Dawson, Daniel, Garrett, Mingo, Perry, Ephraim, Abraham, and Joe Wilkes. This young gang added youth and strength to the insurgent band. Between the horses and the new young faces, the rebel army was gaining a second wind.

Continuing east toward New Orleans, the insurgents passed the Red Church, where François Trépagnier would later be buried. Sparing the minister, they swept down the River Road, passing the two-story Destrehan mansion with its bold architecture and imposing presence. Here Jasmin, Chelemagne, and Gros and Petit Lindor joined the insurrection. Jean Noël Destrehan himself had long since fled for the city.

Here, finally, maroons began to join the insurrection. At the plantation of Alexandre Labranche, the longtime maroons Rubin and Coffy left the swamps and joined the revolt. Following Rubin and Coffy's lead, a wave of swamp denizens gave up the security of their wooded retreats to fight with the rebel army.

As the maroons emerged in triumph from the swamps, the planters continued to flee for safety. Alexandre Labranche, who had waited in the swamps until he was assured the slaves had passed, sneaked through the fields and down to the river, where he took a boat to the other side. From there, he fled toward New Orleans in search of safety. He left his loyal slave François "to keep an eye on the situation"—vision, that essential element of slave discipline, was now in the eyes of the slaves themselves. François was in some sense Charles Deslondes' mirror image—a slave driver who chose to command slaves not in service of rebellion and freedom but in service of the status quo and security. François would fight to hold the plantation world together—even as Charles and his men tore at its seams.

At points, the insurgents were not above inflicting their

own punishments on fellow slaves, forcing those who wavered to join them. While several of the slaves on the Trépagnier plantation willingly joined the slave army, others obstinately refused. So Charles, Kook, Quamana, and their allies raised the stakes, threatening to kill any slaves who would not join. The rebels knew that many slaves preferred slavery and security to freedom and death, and to adjust the odds in this complex calculus they threatened violence, too.

As they moved on to New Orleans, the insurgents set fire to the home of the local doctor. Though a doctor might seem an unlikely target, doctors were often hated figures among slaves. Slave masters employed doctors to manage the health of their slaves—a position that put doctors in direct, intimate, and often objectionable relationships with slaves. These slave patients often had very different approaches to medicine and healing, involving herbal medicine and traditional practices with which they felt more comfortable. They were wary of white doctors, who clearly had in mind not their interests but those of the slaveholders. In the pouring rain, burning down a house took a lot of effort. But the slaves were willing to put in the effort to torch the home of the doctor who had violated the most intimate spaces of their bodies with white medicine.

After burning the home of the local doctor, the rebel army arrived at the Meuillion plantation. Here, at the wealthiest and largest plantation on the German Coast, at least thirteen more slaves joined the insurgency. The rebels then laid waste to Meullion's grand home, pillaging and destroying much of the wealth that the planter had accumulated. They also attempted to set fire to the home, but in the words of one planter, the slave Bazile "did alone fight the fire set to the main house" and "alone, prevented them from stealing many of the effects of the late Meuillion." Half Native American (probably Natchez), Bazile might have felt less of a bond with the largely African slave insurgents.

The slaves marched on through the dark and rain. Well after nightfall, they reached Cannes Brûlées, about fifteen miles northwest of New Orleans. On a clear day, the white spires of the New Orleans cathedral and the masts of the ships assembled in the harbor would have been easily visible. There they entered the Kenner and Henderson plantation, one of the hotbeds of insurrection. Harry Kenner, a light-skinned son of a planter father and a slave mother, was one of the original plotters who had met at the home of Manuel Andry—according to other slaves, one of the "most outstanding brigands." Harry garnered the support of over a dozen men from his plantation. Five men whom owners described variously as carters or plowmen—Peter, Croaker, Smillet, Nontoun, and Charles—laid down their tools and joined the fight. A set of skilled laborers also chose to side with the rebels. Elisha, a driver on the plantation, enlisted, as did the blacksmith Jerry, the hostler Major, the coachman Joseph, and the skilled sugar hand, Harry. Guiam, also a coachman and sugar worker, appropriated one of his owner's horses and, armed with a saber, led all the black males on the plantation toward the nearby home of Cadet Fortier. Lindor, a coachman and carter, assisted the organization of this new charge, acting as the group's drummer.

By this point, the band of slaves had traveled twenty-one miles, a march that would have taken probably seven to ten hours. Documentary evidence links 124 individual slaves to the revolt, while eyewitness observers estimated their numbers at between 200 and 500—rivaling the size of the American military force in the region. The rebel army was now composed almost entirely of young men between twenty and thirty who had been employed as unskilled or low-skilled workers. These men had accomplished much on the first day of the insurrection. They had set fire to the houses of Pierre Reine and Mr. Laclaverie, and killed François Trépagnier and the son of Manuel Andry. They had driven their masters into hiding and seized control of the plantations that had been the sites of their labor and captivity.

Despite their impressive numbers, some guns, and horses, the slave army was not well armed. According to later accounts, "only one half of them were armed with bullets and fusils, and the others with sabers and cane knives." Without proper weapons or means of fighting, the slaves could be outmatched by a small group of well-armed men. However, the fear the slaves had engendered among the planters had been enough to drive the planters from their homes and send them into flight. But intimidation and rumor would only go so far. While the slave-rebels' march had thus far met with little resistance, the white planters had been mobilizing, collecting force, and preparing for a counterattack that would strike that night.

* * *

Traveling through the night, the detachment of troops from New Orleans approached the plantation of Jacques Fortier around four in the morning. Soon thereafter, in the pitch-dark night, the planters discovered the slave army. "The Brigands had posted themselves within a strong picket fence, having also the advantage of two strong brick building belonging to Colonel Fortier's Sugar works," Wade Hampton reported. The slaves seemed to have picked a good spot to defend, well fortified and close to the city. Hampton and the planters met to arrange a plan of attack, delaying any abrupt move in favor of a well-organized strike. They knew what was at stake. "The order of attack was formed the moment the troops reached the ground, and the Infantry & Seamen so disposed as to enclose by a forward movement three Sides of the small enclosure which embraced the buildings, and the Horse at the first signal was to charge the other," Hampton later wrote.

The infantry and the seamen crept into position; the horsemen steeled themselves for a bold cavalry charge. At the crucial moment, Hampton ordered them to attack. Horses galloped, guns fired, and soldiers shouted in the night. But no enemy returned fire.

As the soldiers penetrated the walls and fences and began to search the buildings, they found only a few unarmed slaves. The bulk of the slave army had retreated. In his report to headquarters, Hampton blamed "a few young men who had advanced so near as to discharge their pieces at them" for alarming the slaves before the troops could attack. He speculated that the slaves "were therefore upon the alert, and as the line advanced to encompass them, retired in great silence." More likely, the slaves had left well before the military arrived, alerted not by antsy young white men but by slave spies. Though they found no rebels, the militia found ample evidence of the slaves' presence. Fortier's plantation showed evidence that the slaves had been there, "killing poultry, cooking, eating, drinking, and rioting."

The planters did not know it, but they had fallen for a classic West African military ruse. Warfare practiced in the Kongo especially featured frequent advances and retreats intended to confuse the enemy. The Kongolese soldiers would watch their enemies carefully, waiting for the opportune moment to attack. This strategy allowed them to use their greater numbers to overwhelm better-armed forces.

The slaves' ploy had worked marvelously. Hampton and his men and their horses were too tired to pursue the fugitives farther. And so as the slave army retreated into plantation territory, the American military force took a break at Jacques Fortier's plantation.

Fooled by Charles Deslondes and the slave-rebels, the white army now faced a bleak prospect. As they marched farther into the German Coast, each plantation, each grove of cypresses, could shelter the slave army; and every black slave on every plantation was a potential spy or recruit. The soldiers did not know the terrain very well, and the population was clearly hostile. Any optimism Hampton might have had outside the Fortier plantation faded quickly as these thoughts ran through his head.

The slave army, meanwhile, was marching back upriver. They made good time, traveling about fifteen miles northwest from the Fortier plantation toward the plantation of Bernard Bernoudy. As they navigated the terrain, the slaves planned their strategy for defeating the army resting at the Fortier estate. Their chances of success seemed high—the American forces had fallen for an obvious trap and were now too exhausted to pursue them further. But perhaps they should have been thinking of something else. The slaves had killed Manuel Andry's son, but they had allowed Andry himself to escape. They would come to regret that fateful decision. The ebb and flow of power was about to shift.

January 10, 1811

Eleven

THE BATTLE

After Manuel Andry escaped from Charles Deslondes and his fellow rebels the night before, he had fled to the levee and taken a pirogue across to the southeastern bank of the river. The revolt had not spread across the half-mile-wide Mississippi, and a bleeding Andry headed straight for the plantation home of Charles Perret. Perret and his family owned several plantations stretching around the river bend across from Andry's home. Perret and his family must have been shocked at the sudden apparition of a bleeding and half-dead Manuel Andry. The desperate and furious man had just watched his son murdered by his own slaves, and he had been helpless to do anything but run for his life.

The Perrets listened with shock and horror to Andry's alarming reports—their worst nightmare had become reality. "My poor son has been ferociously murdered by a hord of brigands," Andry reported, who "have committed every kind of mischeif and excesses, which can be expected from a gang of atrocious bandittis of that nature." Andry was not sure exactly what atrocities the so-called brigands might by now have completed, but after watching them kill his son, he expected the worst. And so did the Perrets.

But Charles Perret was a clear-headed young man. He knew what to do. As his family bandaged Andry's wounds, Perret set off on horseback to alert his neighbors. Within hours, the males of all the major families—the d'Arensbourgs, the St. Martins, the Hotards, the Zamoras, the Rixners, the Troxlers, the Dorvins, and the Delerys—assembled together in conference. They knew their livelihood, their way of life, was at risk. They did not trust the American military to do the job, and they were livid at the story of Gilbert Andry's death. The now-bandaged wounds on Andry's body served as a vivid reminder of the consequences of slave revolt.

Watching their own slaves carefully, the planters decided to cross the river and risk their lives to attack the rebel army and "halt the progress of the revolt." They were outnumbered, but, like François Trépagnier before them, they were willing to take a gamble.

Under Perret and Andry's command, eighty planters armed themselves to the teeth and assembled on the levee. With the slave army nowhere to be seen, the planters packed themselves into small pirogues and paddled as quickly as they could across the turbulent and blustery river, navigating the Mississippi's fast currents with a deftness born of long experience. By the time they arrived at the other side of the river, it was about eight in the morning.

The small force marched downriver, hoping to soon encounter the slave army. At about nine in the morning, this second militia discovered the slaves moving by "forced march" toward the high ground on the Bernoudy estate. "We saw the enemy at a very short distance, numbering about 200 men, as many mounted as on foot," wrote Perret.

The planters had unwittingly flanked the slave army. Expecting the only resistance to emerge from New Orleans, the slaves had not anticipated such a rearguard action. They had neither taken a defensive position nor steeled themselves for combat. The planters led by Perret and Andry had come upon them by surprise.

Though the rebels formed a force more than double the size of his detachment, Perret decided to attack. If his planter militia could not defeat the slave army, he believed all would be lost. The planters' wives and children would die at the hands of a ferocious slave army, and everything they had worked to build would burn at the torch. It was a dire thought. Summoning his men, Perret called, "Let those who are willing follow me, and let's move out!" He spurred his men in a forward march toward the slave army.

The slave army slowly formed into a line of battle. They then waited, watching as the planter militia approached slowly across the cane fields. The muskets the slaves carried were only accurate at short distances, so proper military tactics dictated that both sides must close to within 500 feet before firing a single shot. If the slaves discharged their weapons before the planters came that close, they would lose any hope of hitting the target. "In action not one shot out of 100 hit an extended object as high as the head of a horse, at three hundred feet distance," read an 1814 U.S. infantry manual. Fighting an effective battle meant waiting until the enemy was incredibly close, dangerously close, before firing.

We can never know the thoughts that went through the slaves' heads as they took their stand. The two options before them were freedom or death. Fifty years later, a free black man fighting with a regiment of French-speaking ex-slaves from Louisiana described their emotions upon entering battle with the Union army. "We are now fighting, and ask no more glorious death than to die for [freedom]," he wrote. "But for our race to go back into bondage again, to be hunted by dogs through the swamps and cane-brakes, to be set up on the block and sold for gold and silver . . . no never, gladly we would die first." Most likely, the slave-rebels felt the same way.

Both sides faced advantages and disadvantages. The geography favored the white planters. The cane fields formed a wide, flat, open space with good visibility. Most battles in North America were fought in a mix of woods and open space—lending an advantage to guerilla tactics like those the slaves would likely employ. But an open field favored the sort of large-scale infantry movements popular among well-drilled armies. The weather, however, favored the slaves. The pouring rain meant that the white militia had been unable to bring in artillery, either by river or along the muddy River Road. The white planters and their American military allies would not have the benefits of grapeshot or cannon fire as they attacked the slave army. Both sides would fight with muskets.

Staring into the face of death, the slave army did not blink. "The blacks were not intimidated by this army and formed themselves in line," wrote a Spanish agent in New Orleans. Then, in an instant, the first shots rang out. *Recover arms, open pan, handle cartridge, prime, shut pan, load, draw ramrod, ram down, return ramrod, make ready, aim, fire,* went the soldiers with their muskets.

The African drums beat war rhythms and the leaders called out to the slaves to encourage them. As the first soldiers on both sides fired their muskets, clouds of smoke would have quickly poured down on them, hiding everything but the flash of enemy guns.

Guns roared. Muskets crashed and burst. Bullets zipped and hissed through the air. The slaves could only have felt the unease and terror of confronting a danger that they could neither see nor comprehend. The slaves at first might not have recognized the noise of bullets, which could sound like fast-moving bees or birds. Amid the smoke and chaos, men began to drop. Their deaths would have seemed strangely disconnected from the cacophony of noise: the bullets themselves were invisible. The leaders of the slave army watched through the smoke as their men began to fall, as bullets opened gaping wounds in the bulging muscles of the sugar workers.

Perhaps the slaves discharged their weapons too early; perhaps the white planters were simply better trained and disciplined in modern warfare. But within a few minutes, the slaves had discharged all of their ammunition—and the planters kept firing. The slaves watched as corpses proliferated. Their hair still wet from the recent rain, rebel slaves lay dead on the ground, their eyes glazed, their lips blue, and their last expressions fixed forever in their faces. It was a horrifying sight.

A Spanish spy reported that the slave line broke as soon as their ammunition began to peter out. And then the planter militia, some on horseback, charged into the fray as the slaves ran for the relative protection of the cypress swamp. Angered white soldiers poured over the slave lines, shooting and bayoneting the slaves who put up resistance. "Fifteen or twenty of them were killed and fifty prisoners were taken including three of their leaders with uniforms and epaulets," wrote the Spanish spy in New Orleans. "The rest fled quickly into the woods." As the slaves ran for the swamps, they would have heard the desperate cries of the wounded, who knew that they would soon be chopped up by furious white planters. A strange silence settled, pierced only by the shrieks and groans of the wounded. A massacre was under way.

Kook and Quamana were among those captured. These two leaders were ordered sent to the Destrehan plantation, where they would be tried as an example to the other slaves.

The planter Charles Perret glowingly reported that the white forces "left 40 to 45 men on the field of battle, among whom were several chiefs." Most likely the planters killed those survivors immediately after the battle. Only about twenty-five prisoners—including Kook and Quamana—survived to trial. After killing the slaves, the planter militia began a barbaric practice: chopping off the heads of the dead rebels as souvenirs and warnings for other slaves. The blood of the wounded, the dead, and the decapitated soaked into the cane fields. The result, Manuel Andry observed later, was a "considerable slaughter." According to the planters, not a single white soldier fell to the slaves' muskets.

Charles Deslondes was among the slaves who fled to the swamps. Solomon Northup, who worked on a nearby plantation, described the experience of running away into the swamps. "I was desolate, but thankful," he wrote. "Thankful that my life was spared,—desolate and discouraged with the prospect before me."

The fleeing rebels did not have much of a lead. Enlisting the assistance of a party of Native Americans—a strategy that had been used by slaveholders during Louisiana's maroon wars—the militia headed into the swamps, led by packs of bloodhounds trained to chase runaway slaves. "I left with 25 volunteers to beat the bushes, to harass the enemy, and to make contact with those who had fled," Charles Perret reported.

In the silence of the swamps, the escaping slaves could hear the howl of the dogs—dogs they knew to be trained to attack black slaves. "I never knew a slave escaping with his life from Bayou Boeuf," Northup wrote. "In their flight they can go in no direction but a little way without coming to a bayou, when the inevitable alternative is presented, of being drowned or overtaken by the dogs." In the swamps, cypresses gave way to palmetto trees, their heavy leaves darkening the swamps. Moccasin snakes and alligators made the swamps all the more dangerous. Footing was uneven.

At first, the planters found only the wounded, who were unable to run. They discovered one plantation mistress, a Madame Clapion, hiding in the midst of the forests shivering with cold and terror. But before too long, the bloodhounds caught a scent. They did not know it yet, but Charles Deslondes was running just a few hundred feet ahead.

Charles's experience was probably similar to Northup's. The dogs' "long, savage yells announced they were on my track," Northup wrote. "Fear gave me strength, and I exerted it to the utmost." The yelping dogs were gaining, running faster than any man could run. With each howl, Charles would have sensed their imminent approach. "Each moment

I expected they would sink into my back—expected to feel their long teeth sinking into my flesh," wrote Northup. The dogs would tear a man to pieces in minutes, unless called back by their masters—something the enraged planters were unlikely to do.

Charles did not escape. The dogs got to him first, dragging him down and ferociously biting his sweating flesh. The planters, recognizing Charles as "the principal leader of the bandits," brought him back into the cane fields to make a public demonstration. According to one witness, the militiamen chopped off Charles's hands, broke his thighs, shot him dead, and then roasted his remains on a pile of straw. Charles died a martyr for his cause, his death cries a stirring message to the escaped slaves still cowering in the marshes.

* * *

Reprisals continued unabated on Saturday as the militia came upon a band of rebels hiding out in the woods. Flushed out by two detachments of cavalry, the soldiers captured "Pierre Griffe, murderer of M. Thomassin, and Hans Wimprenn, murderer of M. François Trépagnier, and pressed them closely that they came upon M. Deslondes' picket and were killed." The militiamen did more than murder. They hacked off the men's heads and delivered them to the Andry estate.

As the militia hunted down the remaining slaves, federal reinforcements called in by Claiborne converged on the German Coast. Commanding a force of artillery and dragoons,

Major Milton arrived Friday morning from Baton Rouge. In the days prior to the insurrection, Milton had been leading his dragoons north around Lake Pontchartrain in order to fight the Spanish in West Florida. Milton had heard the news at about midday on Thursday, and he had traveled about fifteen miles down the river to the German Coast on an emergency mission to give aid to the militia. Grateful for the extra assistance, Hampton posted Milton and his men in the neighborhood with instructions "to protect and Give Countenance to the Various Companies of the Citizens that are Scouring the Country in Every direction." Hampton concluded that the planters "have had an Opportunity of feeling their physical force [and were] equal to the protection of their own property." Nevertheless, Hampton feared new revolts along the coast, and he ordered Milton to ensure that such insurrections did not occur. "I have Judged it expedient to Order down a Company of L'Artillery and one of Dragoons to Descend from Baton Rouge & to touch at Every Settlement of Consequence, and to Crush any disturbances that May have taken place higher Up."

Hampton was taking no chances, because he did not think the slaves had acted alone. Hampton linked the insurgents with the ongoing war with the Spanish for control of the Gulf. "The [slaves'] plan is unquestionably of Spanish Origin, & has had an extensive Combination," he wrote. "The Chiefs of the party that took the field are both taken, but there is Without doubt Others behind the Curtain Still More formidable." He saw the slave insurrection as a Spanish counterattack on American authority, which was not all that far-fetched.

While Hampton pondered the military and political nature of the uprising, the slaveholders crept out of hiding, called forward by the militia who wanted to secure a familiar kind of peace. Perret ordered the "proprietors to return to their properties" and "all the drivers to carry out the accustomed work at the usual hours." These actions were necessary, the militiaman later explained, "so as to maintain order." For Perret, as for many other slaveholders, "order" meant the reinvigoration of the production of sugar. And so as the planters attempted to pick up the pieces and reestablish that order, they turned to tried-and-true methods of ensuring slave compliance. Only this time, their violence was on a much larger scale.

January 12–21, 1811

Twelve

HEADS ON POLES

THERE IT WAS, BLACK, DRIED, SUNKEN, WITH CLOSED EYELIDS—A HEAD THAT SEEMED TO SLEEP AT THE TOP OF THAT POLE, AND, WITH THE SHRUNKEN DRY LIPS SHOWING A NARROW WHITE LINE OF THE TEETH, WAS SMILING, TOO, SMILING CONTINUOUSLY AT SOME ENDLESS AND JOCOSE DREAM OF THAT ETERNAL SLUMBER. . . THESE HEADS WERE THE HEADS OF REBELS . . . THERE HAD BEEN ENEMIES, CRIMINALS, WORKERS—AND THESE WERE REBELS. THOSE REBELLIOUS HEADS LOOKED VERY SUBDUED TO ME ON THEIR STICKS.

Joseph Conrad

Whether they killed the insurgent slaves immediately upon encountering them, after slow torture, or following a court trial, the planters performed the same spectacular violent ritual. Obsessively, collectively, they chopped off the heads of the slave corpses and put them on display. By the end of January, around 100 dismembered bodies decorated the levee from the Place d'Armes in the center of New Orleans forty miles along the River Road into the heart of the plantation district.

"They were brung here for the sake of their Heads, which decorate our Levee, all the way up the coast," wrote planter Samuel Hambleton. "They look like crows sitting on long poles." Along the course of the revolt, from the plantation of Manuel Andry down the River Road through the gates of New Orleans and into the center of the city, the decomposing heads of slave corpses reminded everyone with a nose, ears, and eyes where power resided. The plantation masters (two of them future U.S. senators), the slaves on the plantations, the boatmen traveling up and down the river, the U.S. military forces in the region, and everyone else who passed through the German Coast in early 1811 traversed a world of rotting bodies. Those passersby no doubt saw themselves in the bodies of the dead, registering, if only subconsciously, the awesome power of a state in the making.

The volunteer militias on the field of battle were the first to practice this ritual—but they only began the bloodletting. From the plantations to the city center, planters, government officials, and military officers—French and American—reenacted the same rite of violence. Ritual, they understood intuitively, imposed coherency, and through coherency, control.

What motivated this savagery? Most likely fear. Commodore Shaw expressed it best. "Had not the most prompt and energetic measures been thus taken, the whole coast would have exhibited a general sense of devastation; every description of property would have been consumed; and the country laid waste by the Rioters," he explained. The public destruction of the rebels was, in slaveholders' minds, a necessary precondition for the safety of the plantation regime and the prevention of a ferocious revolt along the lines of Haiti.

Psychologically, killing another human being is difficult—unless some circumstance makes it possible to dismiss the humanity of the murdered. In this case, a powerful racist ideology that characterized black slaves as little better than cattle, coupled with a rage inspired by a violation of the racial order, provided ample justification. The planters considered the slaves brutal savages hell-bent on wreaking unspeakable atrocities on them and their families. In an area full of planters with strong ties to Haiti, such atrocities were not difficult to imagine.

In committing these atrocities, the planters were using savagery to fight what they understood as savagery. They saw the imagery of heads on pikes as a language that their slaves could understand—corpses represented a lingua franca in interactions between the colonists and the colonized, the masters and the slaves. Planters wanted to make sure that anyone who might empathize with the revolutionaries, anyone who wanted to see the dead as martyrs, would have to reckon with the image of rotting corpses. In the words of Manuel Andry, the planters wanted to "make a GREAT EXAMPLE."

This brutal ceremony was not a new idea, however. The planters were drawing on a long tradition of spectacular violence in service of slavery and colonialism. This was not just a French, African, Spanish, American, Haitian, Native American, or British ritual, but an Atlantic ritual—a ritual that flowed across the trade links that connected America to the

Caribbean and to Africa. In the previous fifty years, beheadings had become the prime method for putting down slave revolts. From 1760 to the early years of the nineteenth century, plots and revolts surfaced with increasing frequency, affecting British, French, Spanish, Dutch, and Danish territories. Both the insurrectionaries and their suppressors used beheadings as a means of discourse. The Coromantee slave Tacky led the first great revolt of this new age of revolution in 1760. The British colonists quickly suppressed his plot, capturing Tacky, decapitating him, and posting his rotting head on a pole. Beheading sanctified the suppression of the uprising. When slaves rebelled in Haiti, the decapitation and public exposure of corpses overwhelmed the island. Both blacks and whites decorated their encampments by hanging corpses from the trees or placing decapitated heads on stakes.

This was not the first time New Orleans had seen such violence, either. In 1795, a group of slaves were tried for conspiring to travel from plantation to plantation, chopping off the planters' heads with their axes. In response to this threat of decapitation by axe (the conspiracy never became realized), the planters struck swiftly and violently. By June 2, the planters had hanged twenty-three slaves and nailed their heads to posts along the Mississippi River from New Orleans to Pointe Coupée. Decapitation and the display of bodies was a well-worn trope of servile insurrections in the Atlantic.

The dishonoring of corpses functioned not only to terrify the slaves but also to reassure white planters of the power of the order they had established. "Condemnation and execution by hanging and beheading are going daily; our citizens appear to be again at ease, and in short, tranquility is in a fair way of being again established," Commodore Shaw wrote. In Shaw's mind, there seemed to be a direct cause-and-effect relationship between conviction, execution, the restoration of order, and the "ease" of the citizens. Witnesses to these spectacles become participants in the restoration of sovereignty, their gazes and their politics co-opted by death, dismemberment, and public decay. At least that is what the planters hoped.

* * *

Kook and Quamana survived the initial bloodletting in the cane fields–turned–killing fields of Bernard Bernoudy. The planters had a special plan for these two men, whose formal military dress no doubt betrayed their leadership role in the uprising. They would try Kook and Quamana, along with several others, in a court of law.

They dragged Kook and Quamana back along the route of their erstwhile uprising, past the haunting site of beheaded corpses, past the burnt mansions and the sites of the previous day's successes, to the heart of the plantation world: the mansion of Jean Noël Destrehan. Perhaps they beat them and tore off their uniforms or perhaps they left them unmolested; few details survive.

When they arrived at the Destrehan plantation, the planters threw Kook and Quamana and nineteen other slaves into a tiny white-washed washhouse just behind the manor. They bolted the wooden door and windows, leaving the slaves huddled on the brick floor of a room barely big enough to fit a table—a cramped space reminiscent of the slave ships that had brought many of them to this world in the first place. The rebels, wrote one planter, were "awaiting the stroke of law, which will be meted out without any kind of delay, especially given the present circumstance, urgent as it is, where it is the question of suppressing a revolt that could assume a serious character, if the chiefs and principal leaders are not promptly destroyed."

As the slaves sat trembling in fear in their makeshift prison, anticipating what brutal tortures might next come their way, Pierre Bauchet St. Martin, the judge of St. Charles Parish, convened a tribunal of slaveholders on Jean Noël Destrehan's plantation. St. Martin fought alongside Andry and Perret—he had been part of the initial bloodbath. The planters—five of them—gathered on the second floor of Destrehan's grand manor, turning the family's ornate parlor into the state's space.

The planters intended their tribunal to legitimize their violence and to help reestablish the boundaries between the civilized and the savage—boundaries ironically blurred by the ritualistic beheadings. They intended the tribunals to swiftly approve the murder of all slaves involved in the uprising so that society could be reestablished to meet the planters' visions. The tribunals were necessary, they explained, "to judge the rebel slaves . . . with the shortest possible delay, particularly in view of the seriousness of the present situation in which it is necessary to suppress a revolt which could take on a ferocious character if the chiefs and principal accomplices are not promptly destroyed."

These trials were not meant for the benefit of the slaves,

but rather to present the powerful as legitimately, ethically, and rightfully powerful. They sought to legitimize death by refracting it through the language of legality. While not particularly interested in the slaves' side of the story, the tribunals nevertheless began their work by interrogating the surviving captives. As Manuel Andry had put it days before, the planters "perfectly knew" who the culprits were. The planters needed from their prisoners only admissions of guilt and assertions of the guilt of others. In some senses, the answers the slaves gave were irrelevant; the only purpose of the questioning was as the preamble to a trial whose end was clear from the beginning: the quick execution of all slaves involved in the insurrection.

Back in the washhouse, the slaves remained literally in the dark about what was going on outside. But before long, they heard the bolts pushed from the door and the planters entered. They wanted Cupidon first. Grabbing him from among his fellow captives, they marched him along the path, up the stairs, and into the parlor where they had assembled.

The planters wanted to know one thing: who was guilty and who was not. The planters left no details about how they treated the captured rebels, but the records of other slave tribunals leave little doubt. "The confessions were extracted by means calculated to excite *the fear of present death in the firmest mind*," read one such trial, held for insurgent slaves in South Carolina. "The prisoners were in irons . . . One strikes [a slave captive] in the face, and threatens to kill him 'if he don't tell all about it'; another says to [another captive] 'tell about it, *they* will hang you if you don't,' and there 'they' stood—an infuriated crowd."

Whatever they did, they made Cupidon talk. As he spoke of the events of the last few days, he began to denounce his fellow rebels. Charles Deslondes, he began, was "principal chief of the brigands." He also denounced three other slaves from the Andry plantation, one, named Zenon, as a "principal brigand." The slave Mathurin, he claimed, had "commanded, armed with a saber." Cupidon confessed that Harry of the Kenner and Henderson estate also played a key role in the uprising. Before long, Cupidon began to talk even of the other slaves locked in the washhouse. He told the assembled planters that "the black Koock (owned by Mr. James Brown)" had "struck Mr. François Trépagnier with an axe . . . leaving him for dead."

The slave Dagobert was next. He too confessed to participating in the revolt and fingered several other slaves as participants—including Cupidon. And he added to the charges against Kook. Kook, he told the assembled planters, had "set fire to the house of Mr. Laclaverie as well as to that of Mr. Reine, the older."

The final interview of the day was with Harry of the Kenner estate, who had been one of those who met with Charles on the Sunday before the revolt. Harry, unlike Cupidon and Dagobert, refused to speak or to confess. The planters sentenced him guilty on the spot. With the first few interrogations complete, the planters took their rest; they would begin the trials again the next day.

On the morning of January 14, after several other slaves had been interrogated, the planters called next for Kook. Kook refused to denounce any other slaves, and he would not tell the planters who participated and who did not. But he did proudly confess to one thing—"he admitted that he was the one who struck Mr. François Trépagnier with an axe." Kook had confirmed his own death sentence. Later in the day, Quamana too would take the same course. "Quamana acknowledged guilt and that he had figured in a notable manner in the insurrection. He did not denounce anyone," read the transcript. The two warriors stayed true to their oaths.

Marched from the makeshift jail to the parlor of the Destrehan manor, many of the accused conformed their words to their owners' scripts. As the slaveholders began their interrogations, they confronted an overwhelming multiplicity of stories. The politics of the washhouse was every bit as complicated as the politics of the mansion.

Some slaves, like Dagobert and Cupidon, were free in their denunciations—and the planters acted on their words. Dagobert, for example, denounced nine of the slaves in the washhouse, all of whom the tribunal later found guilty and sentenced to death. Cupidon denounced ten of his fellow slaves, six of whom would likewise die. But as Cupidon's case suggests, the tribunal weighed rebels' words, and for reasons they kept to themselves, the court refused to act on a number of the slaves' accusations. For example, when Eugene of the Labranche plantation denounced eighteen slaves, the planters sentenced to death only seven of the men he named. The tribunal evidently found Louis of the Trépagnier estate and Gros Lindor of the Destrehan place even less trustworthy, with planters executing only six of the twenty-six men Louis named and two of the fifteen identified by Gros Lindor. The record reveals nothing about why Louis, Gros Lindor, and others denounced the men they did. But it is clear that the planters shared a different opinion.

Other slaves refused to testify or submit to the juridical power of the planters. Robaine, of James Brown's plantation, refused to accuse anyone. Joseph, of the Trépagnier estate, "confessed his guilt and did not deny the charges made against him. He did not accuse anyone." Étienne and Nede of the Trask estate did the same. Amar, of the Charbonnet estate, "did not respond to any of the questions that were addressed to him because he had been wounded in the throat such that he had lost the ability to speak." He had no choice but to remain silent.

The planters made no effort to distinguish between these slaves or to define their crimes on an individual basis. They simply categorized eighteen of the twenty-one slaves as guilty and dismissed entirely the diversity of the slaves' testimony. "These rebels testified against one another, charging one another with capital crimes such as rebellion, assassination, arson, pillaging, etc., etc., etc.," they concluded dismissively.

But beneath this façade of simplicity lay a much richer story: the story of the uprising from the slaves' perspective. During the interrogations, the slaves identified eleven separate leaders. These leaders came from Louisiana, from the Kongo, from the Asante kingdom, and even from white fathers. Their names were French, German, Spanish, West African, and Anglo-American. The politics of the slave quarters were complex and Atlantic. There was no single ideology, nor one single leader, that defined the insurgents or their agenda—rather, the slaves counted in their ranks men from such revolutionary hotbeds as the Kongo, Haiti, and the Louisiana maroon colonies. But amid this chaos, the planters cared only to assign the descriptor "guilty."

Justified by legal proceedings, the planters turned again to violence. Prepared to make the "GREAT EXAMPLE" favored by Andry, the tribunal announced that "in accordance with the authority conferred upon it by the law," it "CONDEMN[ED] TO DEATH, without qualification," eighteen enslaved rebels. Most horrifying was the next line. "The heads of the executed shall be cut off and placed atop a pole on the spot where all can see the punishment meted out for such crimes, also as a terrible example to all who would disturb the public tranquility in the future," read the conclusion of the court.

Their judicial proceedings complete, the planters shot each of the eighteen slaves sentenced to death, and chopped off their heads and put them on pikes. These pikes they drove into the ground of the levee, "where every guilty one will undergo the just chastisement for their crimes, with the end of providing a terrible example to all the malefactors who in the future would seek to disrupt the public tranquility." Kook's and Quamana's heads would be eaten by the crows as the planters returned to their labor.

* * *

In New Orleans, the planters convened a second set of trials, with the same purpose of establishing order through death. The first floor of the City Hall housed not only the guardhouse of the city guard but a prison for runaways (known as the Calaboose, from *calabozo*, the Spanish term for a vaulted dungeon). Here the captured rebel slaves were kept in irons. Some of them probably had been here before. Masters who did not want to punish their slaves could send them to City Hall, where a government official would press the slave flat on his face, binding his hands and feet to four posts before flogging the wretched man the set number of lashes. On the upper floors of the building the magistrates had their offices and courtrooms. Before their chambers, a gallery ran along the whole length of the building, with large windows airing the courtrooms where the slaves were brought for trial. Many citizens packed the courthouse to watch the trials.

Commodore Shaw was among them. "It is presumed that but few of those who have been taken will be acquitted," he wrote as the trials unfolded in the city. Shaw was right. Only a few of those brought before the St. Louis court enjoyed their judges' mercy. Among the favored was thirteen-year-old Jean, the slave of Madame Christien. Though Jean was found guilty of insurrection, his sentence was not death but rather to witness first the death of another slave and then to suffer thirty lashes at the hands of a public official. The court treated Gilbert with leniency, too, but his case turned on his uncle's decision to deliver him to justice and beg for mercy from the court. The court commuted the sentence of Theodore of the Trouard estate because he gave the court valuable information on the recent insurrection.

Gilbert, Jean, and Theodore were exceptions. The New Orleans court sentenced most of the captives to death, ordering their bodies prominently displayed in public places. Within three days of their executions, the remains of John, Hector, Jerry, and Jessamine swayed on the levees in front of their masters' plantations. Étienne and Cesar were "hung at the usual place in the City of New Orleans." Daniel too, at least until his severed head was relocated to the lower gates of the city. Regardless of where their bodies came to rest, the sight and stench of the men's dead flesh bore witness to American—and slaveholder—might.

* * *

It would require more than a hundred rotting bodies, however, to transform Louisiana into a cohesive part of the American union, and Governor Claiborne knew it. Seeing in the event and its grisly aftermath an opportunity to solidify his as well as the nation's control over a new territory, he quickly dismissed both Wade Hampton's belief that the Spanish were to blame for the uprising and the French residents' fear that the rebellion had been a "miniature representation" of Haiti. Instead, and repeatedly—in newspaper columns, private correspondence, and official reports to Washington officials—Claiborne stripped the rebellion of revolutionary or geopolitical meaning by dismissing it as an act of base criminality. Refusing to cede to the slaves what from other perspectives and through other eyes might appear as a deeply political act, Claiborne used the events of January 8 to 11 to dramatize American civil and institutional power, portraying himself as an effective governor and representative of federal authority.

Claiborne worked hard to push a narrative of criminality. In a letter to Jean Noël Destrehan, for example, he repeatedly invoked legal language as he endorsed the planters' spectacular violence. "It is just and I believe absolutely essential to our safety that a proper and great example should be made of the guilty." Claiborne conflated justice and safety. His language functioned to turn rebel slaves into "the guilty," even as Destrehan's violence turned those "guilty" into "examples." The construction of criminality in opposition to law functioned to assert the ubiquity and strength of American government just as it legitimized extreme violence. Claiborne sought first to criminalize, then to marginalize, the potentially revolutionary actions of the slaves. He sought to downplay the power of the insurrection, diminishing it to "mischief." He wrote that "only" two citizens were killed, and that the major harm to the planters came from the depletion of the workforce caused by the large number of slaves killed or executed.

The court trials and Claiborne's representations of spectacular violence as having been meted out only to the guilty served to reinforce a narrative of American control over the Orleans Territory—a narrative supported by the recent conquest of West Florida. In the nineteenth century, courts and legal jurisdiction represented the prime manifestation of American power and national identity. "There can be no stronger evidence of the possession of a country than the free and uncontrolled exercise of jurisdiction within it," wrote a British judge describing the American system of imperial expansion. The court system of the Orleans Territory was a system of political power that served to define and make legible the actions of people, projecting a structure of laws onto the functioning of the body politic. By writing about criminality and brigandage, Claiborne was able to spin the military victory of the planters into a political victory—even though he had played little to no role in the suppression of the uprising.

Not everyone agreed with Claiborne's narrative, however. Anglo-American citizens in New Orleans and elsewhere drew a firm line between the planters' violence and the functioning of the American legal system. A total of twenty-one newspapers, many of them in Pennsylvania, Ohio, and New York, reprinted a January 14 comment from the Louisiana *Courier* that condemned the spectacular violence of the planters. "We are sorry to learn that ferocious sanguinary disposition marked the character of some of the inhabitants. Civilized man ought to remember well his standing, and never let himself sink down to the level of savage; our laws are summary enough and let them govern." The newspaper editors who printed and reprinted this statement drew a firm line between civilization and savagery, condemning this violence as a regression from a state of civilization. But despite this opposition, Claiborne's narrative prevailed where it counted most, among the powerful elite who governed Louisiana and the nation—and, in the centuries to follow, among historians.

Claiborne's portrait of crime and punishment resonated with many of America's political leaders. The news of the insurrection, derived largely from Claiborne's reports, was greeted in Washington with no concerns about the brutality of the suppression. The *National Intelligencer* reported the story on February 19 as almost a nonevent. The paper emphasized that "no doubt exists of their total subdual," referring to the slave insurgents, whom the paper labeled as entirely defeated. The only important element of the story was that the slaves had lost and the planters had won, with the support of the United States Army. Most significantly, Claiborne prevented the revolt from becoming a topic of formal debate in the Senate or the House of Representatives—even though the legislature devoted several weeks to the topic of Louisiana statehood at just around this time. Claiborne succeeded in preventing the uprising from becoming part of the larger political discourse—and in doing so laid the groundwork for the collective amnesia about the 1811 uprising in historical and popular memory.

The German Coast uprising had raised serious questions in the Orleans Territory about the strength of American power, the extent of the Spanish threat, the possibility of a Haitian-style revolution on American soil, and the character of America's newly acquired citizens. The planters realized the urgency of these questions and answered them with 100 dismembered corpses and a set of show trials intended to speak to the local slave population and to all who passed along the Mississippi River. Claiborne spoke to a much larger audience, however, as he represented the main channel of communication to the lawmakers of Washington, D.C. In his reports and published letters, Claiborne took responsibility for the actions of the planters, telling a story of the suppression of the uprising that emphasized the flexing of American military muscle; he wrote the Spanish into oblivion, and excluded the slaves from any sort of political discourse. The governor of a territory whose statehood was being discussed at that very moment in Congress, Claiborne felt the necessity of trivializing the slaves' actions and exaggerating a narrative of government control. Claiborne believed firmly that such violence would forever remain a footnote in the face of a grander narrative of the successes of American empire—

that the end of promoting American authority justified most any means.

If heads on poles were symbols of American authority, they were also symbols of the costs of Americanization. If heads on poles were symbols of control, they were also symbols of the ritual violence that was the constant underlying element of Louisiana society. This was the world Claiborne and the planters made. This was the world they sought to integrate into America. This was New Orleans, and the German Coast, in 1811: a land of death; a land of spectacular violence; a land of sugar, slaves, and violent visions.

January 29, 1811

Thirteen

FRIENDS OF NECESSITY

> THE UNITED STATES SHALL GUARANTEE TO EVERY STATE IN THIS UNION A REPUBLICAN FORM OF GOVERNMENT, AND SHALL PROTECT EACH OF THEM AGAINST INVASION; AND ON APPLICATION OF THE LEGISLATURE, OR OF THE EXECUTIVE (WHEN THE LEGISLATURE CANNOT BE CONVENED) AGAINST DOMESTIC VIOLENCE.
>
> *United States Constitution, Article IV, Section 4*

From 1804 to 1811, Claiborne fought a long, hard, patient struggle to Americanize Louisiana—one that never seemed to gain much traction with the French aristocrats of New Orleans. But in the weeks following the 1811 revolt, Claiborne was able to take a new tone with the sugar masters. He informed them that he would not have time to travel out to the German Coast. He would not survey the burned buildings or see the heads on pikes. He would not stop in at the planters' manors to reassure them. He knew that he did not have to.

Instead, the planters came to him. The once-condescending French aristocrats got in their carriages and, one by one, came into the city to the Cabildo, the seat of American power. They had come to express gratitude and beg for favors—a new position for the German Coast's wealthy elite.

On Tuesday, three weeks after the revolt, both houses of the legislative body of the Orleans Territory filed into the Cabildo to hear Claiborne speak. Americans, Frenchmen, and Spaniards in their most formal clothes sat shoulder to shoulder in the high-ceilinged main hall. Overlooking the Place d'Armes, where slave corpses still rotted in the air, the Cabildo boasted an impressive set of attendees, from Jean Noël Destrehan, the president of the legislative council, to James Brown, the former territorial secretary. The men whispered to each other, eager to hear what Claiborne would say to reassure the residents of their safety.

Calling the assembly to order, Claiborne stood erect with a look of calm and self-possession—even certitude—on his face. By all accounts, Claiborne could be a moving speaker. "It seemed to be a spontaneous effort," wrote a listener to another of Claiborne's speeches. "It had passion and feeling in every sentence, but it was the passion of the heart; satisfied he was right, he was bent on the conviction of others."

All present expected Claiborne to address what was on everyone's mind—the German Coast revolt—but Claiborne began instead by tendering his "warmest congratulations" to those present on the addition of large stretches of West Florida to American control. He proudly sketched out the extent of the new possessions, and he encouraged the residents of New Orleans to make the new citizens feel at home and to facilitate the imposition of American authority. Though he did not mention it, Claiborne believed American expansion should extend beyond Florida into the Caribbean and Central and South America, as well as to British possessions in the Caribbean and Canada. The war in West Florida had not been popular with the French planters, but they now sat silently acceptant, for they were starting to realize the value of a strong American presence in the region.

Finally Claiborne turned to the German Coast revolt. "The late daring and unfortunate Insurrection," Claiborne said as the planters sat up straighter in their chairs, "does not appear to have been of extensive combination; but the result only of previous concert between the slaves of a few neighboring plantations."

Claiborne's attempt to minimize the scope of the revolt did not fool Destrehan and the other planters who had confronted it themselves. They knew that the revolt had been highly organized, extending across plantations thirty miles apart and mobilizing between 200 and 500 slaves. They had interviewed the leaders and put their heads on poles.

Never giving any recognition to the strength and organization of the rebels, Claiborne focused solely on the heroic role white men had played in protecting and defending the city. In particular, he praised the militia and the volunteers who, he said, "made an impression, that will not for a length of time be effaced." The white elite had reestablished their supremacy and beaten back their worst nightmares. This was not a time for questioning, but a time for rejoicing, Claiborne emphasized. As far as the planters and merchants of New Orleans were concerned, Claiborne was right.

In the wake of the revolt and even amid legislative discussions, no government official, legislator, planter, or merchant ever publicly expressed any doubts about the institution of slavery itself. Unlike Virginians after the Nat Turner uprising, the citizens of the Orleans Territory held no debates about emancipation or colonization. Slavery was simply an unquestionable fact of life, no more controversial than the use of currency. And so, as they described and reacted to the uprising, the white elite focused not on changing the base of their society—slavery—but on strengthening the mechanisms that ordered that society—martial law. And with the main military power in the area being the American government, Claiborne sought to channel a desire for improved security into calls for a more robust, and more American, state—a state secure from the Spanish and from the slaves. In the minds of Claiborne and the planters, the proper response to African American political activity was violent suppression backed by the full force of the U.S. government.

To that end, Claiborne turned his speech to the need to militarize Louisiana society. From the beginning of his administration, Claiborne had sought to promote the militia as an important civic institution. Believing, in the words of the Second Amendment, that a well-regulated militia was "necessary to the security of a free State," Claiborne had tried unsuccessfully to rally the planters into militia service under the American government. The planters preferred their own volunteer corps—the same volunteer corps that had defeated Charles Deslondes and the rebel army.

But Claiborne believed now was the time to push the Frenchmen into finally participating responsibly in the American militia. "I could not avail myself of an occasion as favorable as the present, to renew my entreaties for a more energetic Militia System," he intoned. In particular, Claiborne proposed a new militia law that would set times for muster, increase fines for absenteeism, and give the officers the power to imprison or fine those who disobeyed their orders. Claiborne envisioned the participation of every citizen in this new military body—the arming and militarizing of the entire population. "The faithful Citizens cannot but approve such a course," he warned ominously. "They are aware of the many *casualties, internal and external* to which the Territory is exposed, and must be sensible of the importance of a well-regulated Militia."

From the mayor to the president of the legislative council, the city swelled in support of the increased militarization of society and the empowering of armed forces. In nearly one voice, government officials and slaveholder spokesmen declared, "our Security depends on the order and discipline of the Militia." Even Jean Noël Destrehan agreed, arguing that the "late unfortunate Insurrection among the slaves and the untimely end of some of our fellow Citizens, by the unhallowed hands of the desperadoes, and the loss of property to Individuals . . . proves to us the imperious necessity of a prompt organization and discipline of the Militia." A columnist for the Louisiana *Gazette* wrote that New Orleans had learned an "awful lesson" from the revolt and "the time may not be distant when we shall be called . . . against a more formidable foe than the banditti lately quelled." The militarization happened quickly. The militia—once a largely decorative organization—began to meet weekly to train and organize.

The expansion of the militia was only one component of the military reaction to the 1811 revolt. At Claiborne's request, the territorial legislature submitted a petition to President James Madison on February 11, 1811, asking that a regiment of regular troops be stationed permanently in New Orleans. Given "the state of the population," which lay "scattered over a large extent of country along the river—the situation of this defenseless town—the dangers which we have to dread from external hostilities, and from internal insurrections—the difficulties by which the establishment of a convenient system of militia is attended, and several other weighty considerations," they firmly believed that their future existence depended on federal intervention.

The federal government was receptive. The Union, as it was designed in the Constitution, stood firmly committed to protecting planters from the dangers intrinsic in their slave-based society—in fact, such protection was a foundational element of the United States of America. In 1776, the Declaration of Independence prominently showcased black politics—and not in the section about the rights to life, liberty, and the pursuit of happiness. Rather, Thomas Jefferson's final and most significant grievance with the British government was the British crown's threat to incite a slave revolt if the colonists did not fall in line. "[The King] has excited domestic insurrections amongst us," wrote a fearful and angry Jefferson, using the same language of "domestic insurrection" that the French planters of New Orleans would employ thirty-five years later. And when representatives from the colonies arrived by carriage in Philadelphia to draft the Constitution, the fear of slave revolt again took center stage. The authors of the Constitution agreed to guarantee protection to any state facing such "domestic insurrections"—a promise central to convincing southern states to join the Union.

Slave revolts were not just some vague threat to American government: black people across the nation were constantly speaking about, planning, and in many cases executing revolts that threatened white power centers—though never on the scale of the German Coast uprising—and pushed American politicians to shape their government in very specific ways. The residents of New Orleans could have asked for no better government than the American one when it came to protections of slavery.

Recognizing the necessity of federal support for slave-based agriculture, the Speaker of the House of Representatives gave Claiborne's initiative his full support. "Feeling that our destiny is interwoven with theirs, that a common fate awaits us, we shall cherish the Union with a sincere, cordial, and permanent attachment," he declared. "We shall cling to it as the Ark of safety." The federal government reacted with alacrity, dispatching three additional gunboats to New Orleans in the wake of the revolt. The American government, in the face of slave revolts and the military forces of other empires, was the only force capable of guaranteeing the planters safety—and now, more than ever, they realized and accepted that reality.

* * *

Militarization was not the only response to the revolt, however. Claiborne and the legislature worked together on a set of further reforms meant to stabilize the city and crack down on rebellious slaves. The mayor of New Orleans acted quickly to limit slave liberties and to tax the planters for the dangers posed by their slaves. In late January, the mayor sent a message to the city council asking it to prevent the sale of ammunition to black people, to prevent slaves from renting rooms in the city or occupying dwelling places there, and to prevent the slaves from congregating except at funerals and the Sunday dances. The mayor also asked the council to hold slaveholders more accountable for the behavior of their slaves by levying a tax on the planters' most dangerous property. "I believe, Gentlemen, that in fixing the rates of taxes, you should endeavor to place them in preference upon the male negroes," he announced to the territory's planters. "If there is any danger for the public safety, it is the great number of these negroes [that] are responsible." The mayor wanted to hold the planters liable, at least in part, for the behavior of their slaves. Claiborne disagreed. He knew that he finally had the planters' support and he wanted to solidify that support, not anger the planters by taxing them. He wanted to compensate the planters for their losses and provide them with a financial incentive to support the United States government.

On April 25, the government decided to act on Claiborne's recommendation—spending federal money to compensate planters whose slaves had died in the insurrection. They passed an "Act providing for the payment of slaves killed and executed on account of the late insurrection in this Territory." The act provided $300 per slave killed to each planter, and it also provided one-third of the appraised value of any other property destroyed in the insurrection. The editors of the Louisiana *Gazette*, the same paper responsible for printing Claiborne's letters and declarations, believed the act would have a further effect of promoting social cohesion. If compensation were not offered, the paper feared dire consequences. "[The average resident] will not embody for general defence, he will carefully attend to securing and preserving his own property, and finally will not deliver up his culprit slaves into the hands of justice; the evil arising from such a state of things would be incalculable, and would serve to unhinge the strongest tye that unites society." In the months following the insurrection, planters filed claims for about a third of the slaves lost in the insurrection.

Believing that many of the key rebels were of foreign origin, Claiborne also moved to place restrictions on the importation of slaves—restrictions he had been pursuing since 1803. "It is a fact of notoriety that negroes are of Character the most desperate and conduct the most infamous. Convicts pardoned on condition of transportation, the refuse of jails, are frequently introduced into this territory," Claiborne said in a speech to both houses of the legislative body. "The consequences which from a continuance of this traffic are likely to result may be easily anticipated." This was the closest any white resident of New Orleans came to calling the system of slavery into question—and it went over very poorly with the planters. No action was taken, and the importation of slaves surged over the next few years, buoyed by rising sugar prices and an internal slave trade that brought thousands of slaves from all over the country and smuggled in by pirates raiding

Atlantic slave trade ships headed for Cuba to the markets of New Orleans.

Fear—not some sort of divine mandate—drove American expansion in Louisiana. A need to suppress the black population, and fear of external enemies, pushed Americans to develop a new sense of who and what the country was. The federal government acted to support rogue adventurers and profit-hungry slave masters, allying with those who sought first to conquer and then to farm the American frontier. Economic development through slave-based agriculture was a top priority for the United States government—as the Spanish and the slaves quickly discovered. The story of the new alliances formed after the revolt is a microcosm of the larger definition of colonial America as a slave nation.

Fourteen

STATEHOOD AND THE YOUNG AMERICAN NATION

STRANGE AS IT MAY SEEM, WITHIN PLAIN SIGHT OF THIS SAME HOUSE, LOOKING DOWN FROM ITS COMMANDING HEIGHT UPON IT, WAS THE CAPITOL. THE VOICES OF PATRIOTIC REPRESENTATIVES BOASTING OF FREEDOM AND EQUALITY, AND THE RATTLING OF THE POOR SLAVE'S CHAINS, ALMOST COMMINGLED.

Solomon Northup

In May of 1811, as the bodies of the slave-rebels continued to decompose on the levees, William Claiborne called for a convention to write a constitution that would pave the way for Louisiana statehood. Louisiana statehood, Claiborne wrote, would strengthen the Union by "discouraging foreign intrigues" and "internal discontent." By harnessing the power of the national government, Claiborne believed he could guarantee the security and expansion of the plantation economy against threats from the Spanish, the British, and the slaves themselves. Louisiana's newfound political power would aid those who hoped to make slavery a continental system, spread from the Atlantic to the Pacific across the warm southern regions of North America. The German Coast slave revolt was the last significant challenge to the vision of a slaveholding Deep South spread from Georgia to Texas. Claiborne and the planters' victory cleared the path for the next generation of expansionists.

Northern opposition to Louisiana statehood remained insignificant and trivial—the abolitionists and Free-Soilers had not yet risen to prominence or political power. A few northeastern newspapers and politicians spoke up. "The public will indulge what a grand acquisition the new state of Orleans (lately taken into the bosom of the Union by our good Democrats in Congress) will be to this country," wrote a Massachusetts paper, pointing out that as Congress voted to accept Louisiana into the Union, blacks "were at that identical moment endeavoring to cut the throats of their *white fellow citizens.*" But such protests were muted and sporadic. By and large the country favored Louisiana statehood, for Louisiana statehood was the key to a new and stronger American nation that would spread its imperial tendrils across the continent.

Entering the Union during its first formative years, the new state tipped the nation's balance toward the South, the West, and slavery. Representing the new state of Louisiana, William Claiborne, James Brown, and Jean Noël Destrehan were all, at various times, elected United States senators.

Statehood inaugurated a Deep South boom that lasted until the Civil War. The combined population of Louisiana,

Mississippi, and Alabama expanded from 400,000 in 1820 to almost 2.5 million in 1860, and New Orleans became the second largest port—and largest slave market—in the United States. Between 1776 and 1820, America became a slave country. The slave population of North America tripled between the American Revolution and 1820. As the slave population exploded, Kentucky, Tennessee, Louisiana, Mississippi, Alabama, and Missouri entered the Union. By 1820, the course toward the inescapable conflict had been set.

In quick succession over the next several decades, the American nation would rip and roar across the Southwest, securing its own power, and the power of the slaveholders, from the British, the Spanish, the Native Americans, and the Mexicans. The American colossus would knock down these forces of opposition just as Claiborne and the planters had crushed the rebels of 1811—through violence, conquest, and powerful displays of force.

* * *

Only four years later, the residents of New Orleans found opportunity to test their loyalty—and the effect of the military response to the 1811 uprising. In 1812, the Americans and the British went to war over shipping and trade disputes—as well as a frontier battle over British support for Native American tribes in the Ohio area. In 1813, a British navy flotilla composed of three frigates, three sloops, and ten other vessels made its way from Bermuda to Baltimore. There they landed a force of 2,500 British regulars who began the quick march toward Washington, D.C. The American militia, poorly armed and dramatically less experienced than the British, gathered at Bladensburg, Maryland, attempting to fend off the British army. The battle was a disaster for the Americans. Upon receiving news of the British success, President James Madison fled the capital for Virginia.

The British commanders marched in triumph into Washington, D.C. Down the city's grand avenues the troops paraded, arriving at the President's House before nightfall. There the British commanders ate the supper that had been prepared for Madison—before burning down the mansion, the treasury, and several other public buildings. The militia of the nation's capital had proven incompetent in the face of the British army—and America had suffered an embarrassing defeat.

When British troops landed on the Gulf Coast of Louisiana a year later, few observers believed New Orleans stood a chance. But most observers also did not realize the impact of the 1811 revolt—and the military force the planters employed on a day-to-day basis to keep their slaves in check.

As the British threatened to "liberate" Louisiana and called on the French planters to come to the British side, General Andrew Jackson arrived in New Orleans. Denouncing the "perfidious Britons" and their attempts to rally an "incongruous horde of Indian and negro assassins," Jackson issued a call for the planters instead to support the American government. The British, he wrote, threaten "to prostrate the holy temple of our liberty. Can Louisianans, can Frenchmen, can Americans, ever stoop to be the slaves or allies of Britons?" Jackson demanded that the residents of New Orleans "rally around the Eagle of Columbia, secure it from impending danger, or nobly die in the last ditch in its defense." But it was ultimately not Jackson's words that rallied the planters to the American cause, but rather the fear of slave insurrection by "negro assassins" like Charles Deslondes, Kook, and Quamana.

Claiborne reported to the local populace that "the officer Commanding the English Fleet now on this Coast menaces us (in the course of the winter) with black troops." One American in New Orleans warned James Madison that the British might build "a powerful savage and negro army, joined by the slaves of the country ... [to] carry fire and sword thro' that devoted country." Thinking of 1811, this Cassandra knew fire and sword when he saw it. The British did in reality have a regiment of black troops, and they consciously discussed using them in Louisiana. Reaching out to native tribes and slaves seemed an excellent strategy to upset American control of the Gulf Coast and allow for unchecked British advance. The British had offered freedom to slaves who fought on their side, and some did and won their freedom. And Jean Lafitte's pirates and slave smugglers fought on the United States side, earning pardons from the government.

In January of 1815, over a thousand Louisianans rallied to support Jackson and defend the city. Gathering just downriver of the city in a battlefield amid the cane fields, the militia joined the motley crew Jackson had assembled. On January 8—the four-year anniversary of the great slave revolt—the militia proved its mettle, defeating the most advanced and effective army in the world. The English captain reported that his troops had fallen "like blades of grass beneath the scythe of the mower; brigades dispersed like dust before the whirlwind." In the famous Battle of New Orleans, the Americans won the only real victory of the War of 1812. And the residents of New Orleans proved their loyalty, demonstrating military prowess far superior to the weak detachments that protected Washington, D.C.

* * *

The next step in America's imperial project began just after Andrew Jackson's victory at the Battle of New Orleans. Fear of another German Coast uprising became a major justification for further American imperialism in the parts of Florida still controlled by Spain. When the British evacuated Spanish Florida after the War of 1812, they left behind a well-armed garrison of free black soldiers at a British fort at Prospect Bluff. The fort was essentially a safe haven for refugees fleeing slavery in Georgia and Louisiana, including about 300 black men, women, and children. Against the backdrop of the 1811 revolt, General Jackson saw the presence of these armed free blacks just sixty miles away from the American border as a terrible danger, even though these people had given no indication of aggressive intentions. "I have little doubt of the fact that this fort has been established by some villains for the purpose of rapine and plunder, and that it ought to be blown up, regardless of the ground on which it stands," he wrote to his commander.

In July of 1815, Jackson invaded, sending two gunboats and a battery of cannons to Prospect Bluff. These cannons open fired on the fort, and before long, a heated ball hit the principal magazine and exploded, instantly killing 273 of the occupants within the fort and injuring sixty more. The living blacks were returned to American territory and reenslaved. To Jackson, free black people were necessarily "stolen negroes" and slavery was the only suitable place for them in America.

In fact, Jackson used the fear of slave revolts as one of the justifications for his continuing encroachments on Spanish territory. The Spanish controlled all of present-day Florida and parts of Alabama and Mississippi (other than West Florida, the section Claiborne had conquered several years earlier), and they allowed escaped slaves, as well as Native American tribes, to take shelter in exchange for agreeing to side with the Spanish should war with the Americans come. So Andrew Jackson, in direct violation of international law, began a series of violent military and paramilitary cross-border expeditions. His illegal gallivants culminated in 1818, when his armies stormed through Florida to wipe out the remaining Native American tribes and capture escaped slaves. When the Spanish governor at Pensacola protested that Jackson's invasion was illegal and threatened to expel him from Spanish territory, Jackson simply invaded Pensacola.

President Monroe supported Jackson's efforts, refusing to back down in conversations with Spanish diplomats. In his 1818 State of the Union message, President James Monroe discarded Spain's control over Florida as a relic of the past, a figment of maps and treaties but no longer of reality. He declared that the border between the United States and Spanish Florida was nothing more than "an imaginary line in the woods." Spain's inability to transform Florida's "woods" into agricultural settlements protected by military fortifications proved to Monroe that Spanish control over its colonial possessions was exclusively "imaginary" and should no longer have any effect on the actions of American citizens or government agents. At the time, this was a shocking and bold statement for the young—and still weak—American nation to make about its border with the possession of a European imperial nation. Monroe was dismissing the colonial authority of a major international power by asserting that America's form of economic development—based on Thomas Jefferson's vision of homogenous white agrarian settlements—was the only form of settlement that justified political control over land on the American continent. Slave-based agriculture and political control were, in this view, synonymous.

In 1819, recognizing the precariousness of their hold on Florida, the Spanish decided to give up in the face of overwhelming military threats. The United States signed the Adams-Onis Treaty with Spain, renouncing America's claims to Texas in exchange for Florida. Ten years after Mexico gained independence, it too ratified the Adams-Onis Treaty.

In 1828, Jackson—who earned his reputation on the battlefield of New Orleans in the War of 1812, was the nation's most celebrated killer of Native Americans, known for subjugation of the Creek Indians, his subsequent crushing of the Seminoles, and finally his elimination of the Spanish presence in Florida and conquest of that territory for the United States—was elected to the nation's highest office. As president, Jackson presided over one of the most notorious episodes in American history: the Indian Removal of 1830. Veiling his true purposes with humanitarian and patriarchal language, Jackson used federal troops to force all Native Americans to move west of the Mississippi. Jackson firmly believed that the Native American presence east of the Mississippi was unacceptable and intolerable, and so he removed them, cloaking murder, fraud, and rapine under the name "Indian Removal."

The United States did not let legalities stop its expansion. Slave-owning immigrants from the southern states declared Texas independent in 1836 and requested annexation to the United States. In 1845, after a long debate about the dangers of letting another slave state into the Union, the United States annexed Texas, violating the Adams-Onis Treaty. American delegates simultaneously traveled to Mexico City to offer to purchase California and New Mexico, but negotiations failed. The next year, after troops he had stationed on the Texas-Mexico border were attacked during an illegal excursion into Mexico, President James K. Polk declared war. "The invasion [of Texas by Mexican forces] was threatened solely because Texas had determined, in accordance with a solemn resolution of the Congress of the United States, to annex herself to our Union, and under these circumstances it was plainly our duty to extend our protection over her citizens and soil," wrote Polk in his war message. "Mexico has passed the boundary of the United States, has invaded our territory and shed American blood upon American soil."

American troops crushed the fledgling Mexican army, and in the treaty that ended the war, Mexico ceded to the United States California, Nevada, Utah, and parts of Colorado, Arizona, New Mexico, and Wyoming, and formally acknowledged American possession of Texas.

Louisiana's entrance into the Union thus ushered in an age of expansion and imperial violence. Claiborne, like Thomas

Jefferson, James Madison, and James Monroe, shared a vision for just what America would be and how it would look. All these Virginians shared a commitment to an agrarian republic, an empire for liberty controlled and governed by yeoman farmers. Yet the word "farmers" is perhaps the most twisted word in the American political vocabulary: for Jefferson, as for Claiborne, these farmers were more often large slaveholders than independent freeholders. And so by working to expand America's farms, these men took on the task of expanding America's plantation zone, from its origins in Virginia and the Carolinas eventually all the way to Mexico. From 1803 to 1860, slave owners expanded their hold on the North American continent, churning through new land and bringing slaves from the older states to the newer through a vast new domestic slave trade. New Orleans, perfectly positioned as a gateway to the new Southwest, became the nation's largest hub for slave trading—playing a pivotal role in the expansion of American slavery.

And the American military was there to back up this vision of an Empire for Liberty, systematically eliminating any threats to American power. The rebels of 1811 were just some of the first victims, followed soon after by the massacred refugees of Prospect Bluff. Jackson's Indian Wars of the 1810s, his destruction of Spanish forces in Florida, and finally the Trail of Tears represented the next wave. The Mexican War proved the culmination of this expansive American push. And it is no coincidence that the territories brought into America by Polk in the 1850s provided the spark that lit the tinder of civil war.

Fifteen
THE SLAVES WIN THEIR FREEDOM

The slaves who survived the bloodletting after the 1811 uprising would never forget those turbulent January days. If the estimates are correct, between 100 and 400 of the slave-rebels simply faded away. In addition to those who were killed, some disappeared into the swamps. Others went back into plantation life after the revolt, pretending in front of their white masters that they had never participated in a rebellion. Like Jacob, they would tell their masters, "You know me, I am not capable of such a thing." But their children knew better.

In the relative privacy of the slave quarters, the aging rebels passed down their stories to the next generation. And for the fifty years leading up to the Civil War, these stories served as an inspiration for those trapped in slavery. Charles Deslondes', Kook's, and Quamana's deaths became legendary—a stark reminder that revolution was always a possibility.

But it would take a war—and massive resistance on the

part of African Americans—to separate the American nation from its dependence on slavery and secure the freedom for which the 1811 rebels had fought and died.

* * *

In 1861, the second busiest port in the United States seceded from the Union. Louisiana, led by her queen city of New Orleans, signed the Articles of Confederation and prepared for war. French, German, Polish, and Creole regiments rallied to the Confederate flag, as shipbuilders and factories scrambled to provide them with the warships and weapons needed to fight the more industrialized North. As they received the news of Confederate victories at Bull Run and elsewhere, the wealthy planters and merchants did not predict just how fast and how hard the American military would strike in the Gulf.

By February of 1862, the Union fleet was gathering on a barrier island miles from the mouth of the Mississippi River. David Farragut, the flag officer of the United States Navy, had assembled a powerful attack force: seventeen men-of-war mounting 165 guns, twenty mortar boats, and seven steam gunboats.

Two well-armed Confederate forts—Fort Jackson and Fort St. Philip—stood guard over the lower Mississippi, armed with about 150 guns, fifty fire rafts, and a ragtag band of retrofitted Louisiana vessels. The rebels had created a chained set of rafts to block passage. To get at New Orleans, the federal navy would have to run the gauntlet between these two embankments and also destroy the rafts.

But Farragut was a careful and deliberate man. He sent out small expeditions to tie rags to branches to determine wind patterns and the best positions for gunboats. He fixed chain armor to the sides of the ships to protect the engines, fixed chain cables around the boats, placed sandbags around the guns, smeared river mud over all visible areas of the ship, and whitewashed his guns so that the sailors would be able to easily manipulate them in the night.

On April 18, he ordered his fleet of mortars to move upriver and hide in a bend of the river below the two forts. They then opened fire, raining shells down on the Confederates, who had to guess where the federal gunboats were located. In the first day, Farragut's fleet launched over 1,400 shells, setting fire to the wooden barracks, nearly lighting the forts' magazines, and knocking out seven of Fort Jackson's guns. Farragut divided his boats into three watches of four hours each, with orders to fire 168 times per watch.

For four days the bombardment continued, and the Confederates did not surrender. Farragut decided it was time for a change in tactics. He would simply run his ships past the forts in the cover of night—an incredibly risky tactic.

Around two in the morning, the flag officer raised two red lights into the air. It was the signal, seen through the night mist, to advance. Slipping through a breach in the chain raft, the mortars went first, moving into position near the forts and opening up such heavy fire as to drive the Confederates from their guns.

Then, under the cover of darkness, Farragut's mud-smeared naval fleet began to creep by the forts. Despite Confederate fire, all but three of his boats made it past. And once past the forts, they caught the Louisiana navy by surprise, annihilating every ship with mortar fire and barely receiving any return fire.

Farragut had done what the Confederate government in New Orleans thought impossible: run an entire federal fleet through the twin forts. "People were amazed, and could scarcely realize the awful fact, and ran hither and thither in speechless astonishment," wrote an English observer.

The planters and their slaves lined the riverbank to watch the bold fleet move inexorably toward New Orleans. To the white residents, the sight of Union ships in the heart of the Confederacy was a depressing omen of the collapse of their society. But to the slaves, the ships represented a beacon of freedom. "To the negroes we evidently appear as friends and redeemers," wrote one sailor. "Such joyous gatherings of dark faces, such deep-chested shouts of welcome and deliverance, such a waving of green boughs and white vestments, and even of pickaninnies, such a bending of knees . . . salutes our eyes . . . as makes me grateful to Heaven for this hour of triumph."

These joyous celebrations were only the beginning. The slaves reacted instantly to the presence of the Union army—with armed allies in control of New Orleans, they would not remain slaves for long.

As Union columns advanced into cane country, one writer wrote that it was "like thrusting a walking stick into an ant-hill." As soon as General Benjamin Butler's troops arrived, hundreds of slaves abandoned their plantations and rushed to Union lines. Those who stayed behind refused to work. Plantation discipline collapsed. "Revolt & Insurrection among the negroes," wrote one planter. The slaves "all went up to McManus's plantation. Returned with flags & drums shouting Abe Lincoln and Freedom."

A planter just outside of the city recorded the disintegration of his world in his plantation diary. On July 7, ten of his slaves ran away. The next day, he reported a "stampede" from the plantations toward Union lines. In early October, he reported that the slaves in the neighborhood were in a state of discontent, and he feared many would soon escape to Union lines. Things happened faster than he expected: by the end of the month, the slaves had all left. In early November, he recorded a rebellion on a nearby plantation. The slaves were forcing the disintegration of the plantation economy.

Everywhere across the state, the slaves picked up and left their plantations. In August of 1862, slaves on the German Coast, armed with cane knives, scythe blades, and clubs, marched toward the city to demand their freedom. Eleven succeeded in finding respite among the Union soldiers, several died, and the rest were captured.

Generals on the ground had to deal with the influx of what they referred to as "contraband." "I shall treat the negro with as much tenderness as possible, but I assure you it is quite impossible to free them here and now without a San Domingo. There is no doubt that an insurrection is only prevented by our bayonets," wrote General Butler. "We shall have a negro insurrection here I fancy."

Butler attempted at first to prevent the slaves from entering his lines, but the slaves would hear none of it. Though the Emancipation Proclamation had not yet been signed, these slaves knew that the presence of the Union army meant freedom. The planters could no longer apply the force necessary to keep the slaves in submission, and so the slaves simply up and left. There was no longer any reason to remain in captivity. Before long, Butler began to advocate for some sort of policy that would allow him formally to employ the increasingly large numbers of black escapees who rushed to Union lines. (Notably, Lincoln reprimanded Butler for his actions to emancipate contraband slaves.)

And it was not just in Louisiana where this occurred. Across the front lines of the Civil War, slavery was disintegrating. By the fall of 1862, 200,000 slaves had escaped to Union lines and offered themselves to work for the army in whatever capacity necessary. In virtually all cases, the slaves freed themselves once the Union forces were near. Moreover, the Union army did not seem capable of winning the war without the aid of black men.

Generals and political leaders simply could not avoid emancipating the slaves. "Any attempt to secure peace to the whites while leaving the blacks in chains . . . will be labor lost," wrote Frederick Douglass. "The American people and the Government at Washington may refuse to recognize it for a time; but the 'inexorable logic of events' will force it upon them in the end." Day by day, as the Union fought on—mostly unsuccessfully—against the Confederate army, Douglass's argument seemed to be more and more true.

Lincoln issued a preliminary Emancipation Proclamation in September 1862. The proclamation stated that slavery would be abolished only in areas actively in rebellion—leaving the Confederate states three months to surrender before their black slaves would be turned against them. In January 1863, the proclamation took effect, "in time of actual armed rebellion against authority and government of the United States, and as a fit and necessary war measure for suppressing said rebellion." But the proclamation did not apply to the lands where Charles, Kook, Quamana, and their brethren lived, fought, and died. In a parenthetical statement, the proclamation read, "(except the Parishes of St. Bernard, Plaquemines, Jefferson, St. John, St. Charles, St. James Ascension, Assumption, Terrebonne, Lafourche, St. Mary, St. Martin, and Orleans, including the City of New Orleans)." For these areas were already under federal control and not actively in rebellion.

But the slaves did not let such legalities interfere with their interpretation of the document. In their minds, they were now forever free. At Mooreland Plantation, workers put a Confederate soldier into the plantation stocks, entertaining themselves by mocking and abusing him. The slaves gathered in great convocations to pray and determine their next course of action. Slaves at the Bayou Teche plantation lit a huge bonfire and hundreds danced to the sound of fiddles. Cries of "Glory to God" and "Glory to Abe Lincoln" could be heard ringing out in the night air.

The planters panicked. Whites across the state banded together into paramilitary vigilance committees to attempt to subjugate their newly emancipated labor force. Four months after Lincoln's proclamation, a planter in Grand Coteau wrote that "the slave population is . . . in a state of disorganization and without control committing depredations that menace the honor, the virtue and lives of our families." Men who had built their livelihoods—and their senses of self—on slavery now faced the collapse of their entire world.

But the worst was yet to come. Lincoln's proclamation freed the slaves in all areas not controlled by the Union army and allowed for the enlistment of black troops. The contraband slaves who had fled to Union lines were now free—and free to enlist in the military.

Black men recognized that fighting in the war was a way to secure the rights of citizenship, and service in the military conferred a new sense of dignity and manhood to many former slaves. "If we hadn't become sojers, all might have gone back as it was before; our freedom might have slipped through de two houses of Congress & President Linkum's four years might have passed by & notin been done for we," one black corporal told his men in 1864. "But now tings can never go back, because we have shoed our energy & our courage & our naturally manhood."

For twenty-three-year-old Octave Johnson, joining the Union army meant everything. A planter had sold his mother away when Johnson was fourteen, before selling him for $2,400 six years later. "One morning the bell was rung for us to go to work so early that I could not see, and I lay still, because I was working by task; for this the overseer was going to have me whipped, and I ran away to the woods," he recalled. For years, he lived four miles behind the plantation house, deep in the swamps, stealing beef cattle and sleeping on logs. For months, he was hunted by hounds—eight of which he killed during the course of his maroonage. When Union forces arrived, Johnson saw the perfect opportunity. He escaped to the Union outpost at Camp Parapet and then worked his way up from the commissary's office to a position in one of the first black regiments.

Johnson was one of the first to make the transition from refugee to soldier. The first black regiments were raised in the fall of 1862, after the announcement of the Emancipation Proclamation. The first black regiments came from New Orleans, Missouri, Kansas, and the Sea Islands of south Georgia. Black regiments from Massachusetts, Rhode Island, and Connecticut emerged in 1863. In May of 1863, the War Department established the Bureau of Colored Troops, sending Northern agents to recruit freed slaves from Southern areas held by Union troops.

Initially, there was much resistance to the idea of black troops. White Northerners objected to the enlistment of black soldiers both because they did not want to fight side by side with African Americans and because they suspected African American soldiers of cowardice. "There is not one man in ten but would feel himself degraded as a volunteer if negro equality is to be the order in the field of battle," one New York officer wrote. Black soldiers would have to prove themselves in battle.

Over the next two years, black soldiers proved themselves again and again on the field of battle. Black soldiers totaling 178,985 enlisted men and 7,122 officers served in the war. Of these, 37,300 blacks laid down their lives for freedom. Seventeen black soldiers and four African American sailors won Congressional Medals of Honor. They fought in 449 engagements, of which thirty-nine were major battles.

By the end of the war, black soldiers constituted nearly 10 percent of the fighting force of the North. Lincoln recognized their importance to the Union war effort. "We can not spare the hundred and forty or fifty thousand now serving us as soldiers, seamen, and laborers," Lincoln wrote in 1864. "This is not a question of sentiment or taste, but one of physical force which may be measured and estimated as horse-power and steam-power are measured and estimated. Keep it and you can save the Union. Throw it away, and the Union goes with it." African Americans had made themselves indispensable to the war effort, and they demanded new rights in return.

For the men who served in the military, the experience of fighting against the slaveholders, of becoming liberators, was a powerful and decisive moment. One soldier, Spotswood Rice, wrote to his mistress, a former slave, from his army encampment to warn her of the new social reality. "We are now making up a bout one thoughsand blacke troops to Come up tharough and wont to come through Glasgow and when we come wo be to Copperhood rabbels and to the slaveholding rebels for we don't expect to leave them there root neor branch," he wrote. Rice, like many of his fellow soldiers, took advantage of the presence of the Union army to push for new social and political relations.

The planters took note of this dramatic change in social relations. A white man in St. Bernard Parish wrote to the Union army to complain of the impact of black soldiers on the labor force. He wrote that black members of the First Louisiana Native Guard "visited plantations of loyal & peaceable men, putting guards over their houses, threatening to shoot any white person attempting to leave the houses, and then seizing the horses carts & mules for the purpose of transporting men women & children from the plantations to the city of New Orleans." The black soldiers rallied a group of seventy-five, who, he wrote "went singing, shouting & marauding through the Parish." They even had the temerity to show up to one planter's doors with muskets to demand "some colored women whom they called their wives." The tides had turned, and the bottom rail was now top.

By 1865, black men and women had won themselves a new place in the Union. Half a million slaves had escaped slavery and forced the Union army, and eventually the president of the United States, to declare them legally free. Two hundred thousand black men had fought for the Union army, helping liberate tens of thousands of slaves.

When the Civil War began, no branch of the U.S. government—not the legislature, not the executive, not the judicial—expressed any intention of abolishing slavery. In his first inaugural address, Abraham Lincoln had expressed his support for a constitutional amendment to ensure that "the Federal Government shall never interfere with the domestic institutions of the states." He had, he declared, "no purpose, directly or indirectly, to interfere with the institution of slavery in the States where it exists." The Republican Party, then in control of both houses of Congress, had taken a similar stance. "Never on earth did the Republican Party propose to abolish Slavery," wrote Horace Greeley, a Republican spokesman. "Its object with respect to Slavery is simply, nakedly, avowedly, its restriction to the existing states." In 1857, the U.S. Supreme Court had ruled in the *Dred Scott* case that any attempt to prohibit the spread of slavery was unconstitutional and that African Americans had no right to U.S. citizenship. Chief Justice Robert Taney wrote that blacks "were so far inferior, that they had no rights which the white man was bound to respect; and that [all blacks] might justly and lawfully be reduced to slavery."

Black slaves accomplished their own emancipation through brave and decisive action. The actions of black men in hundreds of military engagements proved integral to the Union war effort, forcing the U.S. government to recognize them as equal citizens. In the two years following the end of the war and black people's exemplary wartime service, the United States passed three constitutional amendments abolishing slavery, establishing black people as citizens, and declaring equality under the law. Fifty years after crows devoured the rotting, decapitated heads of the slave-rebels of 1811, the dream of black political power on the North American continent finally became reality.

Sixteen

THE COVER-UP

The plane cuts low over the flat Mississippi floodplains. From the window, the Mississippi looks brown and placid, passing through the farmland and industrial tracts north of the city. Large tankers and cargo ships seem fixed as the plane swoops in toward Louis Armstrong Airport, which sits on the site of the former Kenner and Henderson plantation.

Though city authorities renamed the airport after a prominent African American, Louis Armstrong, the names of the surrounding towns and streets date back much further. The old River Road sweeps past the poor, primarily black town of Destrehan, before entering the town of Kenner, where the airport is located. The River Road becomes Third Street, then Jefferson Highway, and finally South Claiborne Ave.

Don't bother looking for Charles Deslondes Boulevard or Quamana Avenue. And don't spend much time looking for historical markers of the 1811 revolt. There's only one, across the street from a McDonald's in Norco, nearly forty miles outside of the city center.

Driving along the path of the revolt today, you will pass huge chemical refineries, the sugar plantations of the current day. Overshadowed by these chemical plants, though, the history of older days still survives.

Many residents can trace their ancestry back to sugar slaves from this same area. Slave burial plots, cemeteries full of Civil War–era headstones, and even several remaining plantation homes still keep watch over the German Coast's old ghosts.

The Destrehan plantation is open now for tours—and weddings or parties, if you're interested. A group of prominent white families converted the Destrehan plantation into a museum, seeking to preserve their heritage and remember their own past.

The tour focuses on the lifestyles, family histories, and architectural accomplishments of the planter class. The tour is rich with descriptions of the planters' meals, their parties, and their elaborate family dramas. The architecture is a special emphasis of the tour.

When it comes to slavery, the tour guides describe a system of "Creole slavery" that was generous and fair to the slaves. Slavery was not as bad under the French as it became under the Americans, the tour guides suggest. "Everyone worked, from family members to slaves, because life on a plantation was not easy," reads the plantation brochure. "It has been documented that slaves at Destrehan Plantation were treated with fairness and their health needs provided for."

But even the relatives of Jean Noël Destrehan cannot deny the events of January 1811. In a converted slave cabin not featured on the standard tour, the tour guides have constructed a museum to the 1811 uprising. With brief descriptions of the major events, the cabin features folk paintings that imagine what the event would have looked like. Just as in the history books, the story of slave politics is compartmentalized away from the central narrative of American history.

Though the world of the German Coast seems to have avoided confronting its past, one man has led a group to force New Orleans to do just that. Leon Waters, a sixty-year-old activist who has been involved with radical political causes since the Vietnam War, now provides tours of the uprising to curious student groups and tourists from out of town. "Hidden History Tours provides authentic presentations of history that are not well known," promises Waters's Web site. "We take you to the places, acquaint you with the people, and share their struggles that are rich and varied. These struggles have been made by Africans, African-Americans, Labor and Women. For too long their stories have been kept hush hush. But not anymore!"

A participant in the Black Workers Congress, Waters devoted his early life to organizing factories, even moving to Detroit to head up an effort to create a national struggle against wage slavery. During the past twenty-five years, he has worked for the Afro-American History Society of New Orleans, fighting to restore a "scientific" perspective to the history of the area.

For Waters, the tour represents a way to keep alive the memory of the uprising and the memory of the tradition of "revolutionary struggle" in America. He sees the 1811 uprising as the intellectual antecedent of the American civil rights movement. Aside from delivering tours, he works to fight police brutality and generally promote political discourse among African Americans in New Orleans. Waters has organized several commemorative celebrations of the uprising, featuring marches, reenactments, and speeches.

Though virtually unknown outside of his community, Waters is perhaps the most knowledgeable man in the country about the 1811 revolt. Growing up in the area, he remembered hearing stories from his great-aunt about the revolt. In the early 1990s, he decided to see if there was any truth behind the oral legends of his upbringing.

Waters formed the Afro-American Historical Society of New Orleans, and he set out to uncover history. He and his friends traveled to libraries, courthouses, and archives, where they were often met with obfuscation and racist comments. But they continued their search—and it soon began to bear fruit.

By 1996, they had assembled 168 pages of documents collected from archives all over the country, some of which they had translated from French by a scholar with Haitian roots. Waters arranged for an independent scholar named Albert Thrasher to write up a description of the events, and the group published the results of their research in 1996 at an independent press in New Orleans. The book, *On to New Orleans!*, provides a history of the "revolutionary struggle" of African Americans from 1500 through Reconstruction, devoting twenty-four pages to the uprising and its suppression. In an account defined by Marxist ideology, Thrasher fit the uprising within a long contextual history of revolutionary struggle. The book described the goal of the uprising as to "overthrow their oppressors, to destroy the power of the 'white' rulers."

In addition to compiling a near-authoritative collection of documents related to the revolt, the book drew extensively on oral history from Louisiana. The oral history suggested that the black marchers had two chants, "On to New Orleans" and "Freedom or death," which they shouted as they moved toward the city. Thrasher speculated that Claiborne's account of only two white casualties was "patently in contradiction with the truth" and cited several sources from much later periods that argued for a larger body count.

Thrasher argued that while the revolt was tactically a failure, strategically it was hugely successful. "This revolt stimulated a whole range of revolutionary actions among the African slaves in the U.S.A. in subsequent years," he wrote. "It continued and invigorated the tradition of revolutionary struggle among the African slaves in the Territory of Orleans that would never abate." Though the book is full of Marxist-Leninist language denouncing the "sham" U.S. democracy and the "capitalist moneybags," the book nevertheless provides a substantial account of the event itself.

Few outside of academia and the local community have ever seen the few copies of this book that exist. In fact, there is a vast collective amnesia in New Orleans and the United States more broadly about the massive bloodletting of January 1811.

How did the 1811 uprising become lost in the footnotes of history? How did historians overlook it for 200 years? And why has no one ever bothered to tell the story of the enslaved men who lost their lives fighting for their own freedom? The answers to these questions lie in America's complicated racial politics.

Claiborne wrote the first draft of history—and he wrote it with the goal of covering up the revolt and saving face before an anxious nation. He wrote the slave-rebels out of history, believing that all that was important was the rise of American power in the Southwest. Swallowing Claiborne's interpretation, most historians have portrayed the slave-rebels not as political revolutionaries but as common criminals. Up until World War II, most of these historians advocated or supported the control of white men over the political institutions of the South, conflating the idea of the law with the idea of white supremacy. Examining the revolt from this perspective, they heartily agreed with Claiborne's interpretations and the planters' violent actions.

The Communist movement represented the first challenge to Claiborne's political agenda. After World War II, a wave of activist historians revisited the history of slave revolts in an effort to narrate a history of violent resistance and class struggle that would support their present-day efforts to organize opposition to Jim Crow rule in the South. But while these historians changed the tone of the commentary on slave revolts, they nevertheless kept the basics of the story untouched. Many of these men saw the slaves as mere gears in the great machinery of class struggle, and they saw little need to explore the politics of the enslaved.

* * *

The first historical account of the 1811 uprising emerged amid the political turmoil of Reconstruction, a time when newly emancipated African Americans were agitating for more rights and more control over the terms of their labor. Horrified by this turn of events, a sixty-one-year-old ex-Confederate named Charles Gayarré published an account of the uprising in the final volume of his series on the history of Louisiana. Gayarré believed strongly in the propriety of terror, and the rights of white planters over ex-slaves. "This incident, among many others, shows how little that population is to be dreaded, when confronted by the superior race to whose care Providence has intrusted their protection and gradual civilization," Gayarré wrote in 1866. "The misguided negroes ... had been deluded into this foolish attempt at gaining a position in society, which, for the welfare of their own race, will ever be denied to it in the Southern States of North America, as long as their white population is not annihilated or subjugated." Gayarré, like Claiborne, endorsed the force that the planters used to suppress the uprising. Like Claiborne, he saw planter violence as both necessary and just.

To bolster his argument, Gayarré added an apocryphal story about François Trépagnier. According to Gayarré, Trépagnier heard about the uprising from his slaves but decided to remain on his plantation to protect his property. From here, Gayarré embellished, Trépagnier took a stand on the "high circular gallery which belted his house, and from which he could see at a distance" and "waited calmly the coming of his foes." Trépagnier heard the "Bacchanalian shouts" of the slaves, and he readied himself for battle. "But at the sight of the double-barreled gun which was leveled at them, and which they knew to be in the hands of a most expert shot, they wavered, lacked self-sacrificing devotion to accomplish their end, and finally passed on, after having vented their disappointed wrath in fearful shrieks and demoniacal gesticulations," wrote Gayarré. "Shaking at the planter their fists, and whatever weapons they had, they swore soon to come back for the purpose of cutting his throat. They were about five hundred, and one single man, well armed, kept them at bay." The origins of this story are unclear. Perhaps Gayarré was drawing on oral history, or perhaps he invented this story himself. Perhaps Gayarré had never been out to the Red Church on the German Coast, where Trépagnier is buried beneath a gravestone that reads in French, "François Trépagnier, Killed by Insurgent Slaves on 10 January 1811."

Gayarré was just the first historian to accept uncritically Claiborne and the planters' reading of the revolt. In 1918, prominent Yale historian Ulrich B. Phillips devoted a sentence in his book *American Negro Slavery* to the uprising. Phillips included this sentence in a chapter on "slave crime," pursuing the same narrative of criminality that Claiborne and Andry had so cleverly adopted. Interestingly, Phillips included the Haitian revolution in this chapter as well. The slaves, he wrote, "were largely deprived of that incentive to conformity which under normal conditions the hope of individual advancement so strongly gives." Slaves committed crimes out of backwardness and a lack of civilization, and their "lawbreaking had few distinctive characteristics." Phillips saw revolt as fundamentally apolitical, producing disquiet but little else. Phillips, like Claiborne, saw Southern society as synonymous with white society. Phillips argued that the South was defined by a commitment to racial superiority and to a specific form of social order. "It is a land with unity despite its diversity, with a people having common joys and common sorrows, and, above all, as to the white folk a people with a common resolve indomitably maintained—that it shall be and remain a white man's country," he wrote.

While Phillips only mentioned the uprising in passing, historians in New Orleans wrote more extensive narratives. In 1939, New Orleans journalist-turned-professor John Kendall wrote an article about slavery in Louisiana that depicted black people as casting a "shadow over the city." Kendall argued that the fear inspired by the 1811 revolt was one of the central elements of the New Orleans mentality. In his essay, Kendall wrote a three-page account of the event—an account that remained for many years the most significant and definitive account of the uprising. Kendall depicted the slaves as animals, using words like "growling" and "howling" to describe the "savages" involved with the rebellion. Like Gayarré, Kendall saw this story as a moral tableau. "One must hold the reins tight over the blacks," Kendall wrote. "They must know who were their masters." Rich with overtones about class and race, Kendall's story was meant consciously to generate a certain arrangement of power.

Claiborne wrote the first draft of the history of the uprising; historians like Phillips, Kendall, and Gayarré helped enshrine that draft as the conventional story. Like Claiborne, these men lived in a society where the rule of law and the rule of white men were synonymous. But that vision of society was under pressure. A new movement for racial and political equality was gaining steam through the work of

Communist activists. This movement resonated in academia. The same year that John Kendall published "Shadow over the City," a young academic named Herbert Aptheker joined the Communist Party of the United States. Aptheker had been studying at Columbia University, where he became involved with Marxist efforts to organize Southern tenant farmers into unions. White and Jewish, Aptheker saw two purposes to his alignment with the Communists. According to the *New York Times*, he "saw [Communism] as an anti-fascist force and a progressive voice for race relations." These two motives were fundamentally interconnected. With Hitler gaining power in Germany, many Jews in the United States became worried about the clear overlaps between white supremacy in the South and Nazism. They feared that racist laws could provide a slippery slope into anti-Semitic laws, and that Jewish activism on behalf of African Americans was the best bulwark against the spread of such dangerous ideologies. Aptheker was challenging the Nazis and Jim Crow.

But Aptheker was more than an organizer; he was also an avid writer who, in 1943, published a book that would turn the scholarship of Gayarré, Phillips, and Kendall upside down. His book, *American Negro Slave Revolts*, with a title that consciously imitated Phillips's, forever shattered the myth of the contented slave and forced a reevaluation of the nature of slave revolts. In his introduction, Aptheker attacked Phillips, laying down the gauntlet between Communists and white supremacists. "Ulrich B. Phillips, who is generally considered the outstanding authority on the institution of American Negro slavery, expressed it as his opinion that 'slave revolts and plots very seldom occurred in the United States,'" wrote

Aptheker. "This conclusion coincided with, indeed, was necessary for the maintenance of, Professor Phillips's racialistic notions that led him to describe the Negro as suffering from 'inherited ineptitude,' and as being stupid, negligent, docile, inconstant, dilatory, and 'by racial quality submissive.'" It was a bold move for a twenty-eight-year-old newly minted PhD to attack so openly the established leader in his field, a much-lauded Yale professor, but Aptheker did not refrain from labeling Phillips a racist. However, he devoted little time or attention to the German Coast. Aptheker devoted a short paragraph to the 1811 uprising, describing its size and location, but offered no more details than Phillips. Aptheker's grand political and historical agenda overshadowed the details and the individuals involved.

Tossed from white supremacists to Marxist activists, the actual history of the 1811 uprising fell by the wayside among ideological battles over race and politics. More than fifty years since Aptheker's book was published, the number of historians who have seriously examined the uprising can be counted on one hand. With the notable exception of Hamilton professor Robert Paquette and a few others, most have shied away from attempting to decipher or interpret this complex event—preferring instead to rely on the earlier biased accounts from writers with clear white supremacist or Marxist ideologies. With little attention from scholars, North America's largest antebellum slave revolt has languished in the footnotes of history for 200 years. While historians jostled to write about Nat Turner, who had mobilized fewer than 100 slaves, this diverse band of Louisiana slaves has been remembered only by a few.

But despite its absence from textbooks, the story of the 1811 uprising is central to the history of this country. This is a story about American expansion and the foundations of American authority. Most important, however, this is the story of a revolution organized by enslaved men. These men saw violence as the means to ends they never realized. But their failure to achieve those goals does not mean they did not have goals, or that the sum total of this story was that of the quick and violent suppression of a horde of brigands. Rather, 200 years later, historians must reckon with the politics of the enslaved, with the world the slaves made, and with the humanity of those who fought against slave power. Only through understanding their stories can we begin to comprehend the true history of Louisiana, and with it, the nation.

EPILOGUE

In the summer of 1957, things were heating up in the small town of Monroe, North Carolina. The problem was simple: the only pool in the area was barred to blacks, and several black children had drowned swimming in unsafe swimming holes. The president of the local chapter of the National Association for the Advancement of Colored People, a man named Robert F. Williams, decided that the best solution would be for the pool to desegregate in accordance with the recent *Brown v. Board of Education* Supreme Court decision.

Williams's wife, Mabel, thought the plan was preposterous. "White folks don't want you to sit beside them on the bus, Rob," she said. "You really think they're gonna let you jump in the water with them half-naked?"

Williams insisted that he was going to try. He showed up at the entrance to the pool with a group of eight black children with bathing suits and towels. Turned away, they returned the next day, and the day after, standing outside the gates in protest of an injustice both humiliating and life-threatening to the area's black children.

But most white residents of Monroe did not see the situation the same way. With the recent closing of a textile mill and a growing backlash against the racial progress brought about by World War II, the Ku Klux Klan had made a powerful resurgence in the area. The evangelist and Klan leader James "Catfish" Cole barnstormed through the Carolina piedmont, inciting thousands of white supremacists to a series of rallies, cross burnings, and even dynamite attacks on black activists. Cole saw the pool protests as a perfect platform for his message.

Rallying the Klan on Highway 74 outside Monroe, Cole told 2,000 of his followers, "a nigger who wants to go to a white swimming pool is not looking for a bath. He is looking for a funeral." After burning crosses, the Klan began a series of raids on the black neighborhoods of Monroe. They drove their cars, horns blaring, through the neighborhood, throwing bottles and firing pistols in the air. Cole and his henchmen didn't expect what happened next.

Robert Williams, member of the NAACP and the National Rifle Association, strapped a pistol to his belt and took a stand. Along with a group of other military veterans, Williams dug foxholes and built a rifle range in his neighborhood. "We got our own M-1's and got our own Mausers and German semiautomatic rifles, and steel helmets. We had everything," he later recalled. The female members of the NAACP set up an emergency phone tree, and Williams and his men prepared for battle.

On October 5, 1957, Cole held a huge rally in Monroe. As the rally finished, the Klansmen leapt into their cars, shotguns in hand, ready to have a little fun in the black neighborhood. But as they drove along familiar roads, they saw an unusual site. Behind sandbag fortifications and earth entrenchments, Williams and his men kneeled with guns pointing outward. As the Klan neared, the black citizens let loose a volley of bullets.

What happened next was best described by one of the men behind the entrenchments that night. "When we started firing, they run. [The Klan] hauled it and never did come back." Williams humiliated the Klan—and sent a powerful message to the white community of Monroe. Taken aback by this new turn of events, the largely white city council held an emergency session and passed legislation banning KKK motorcades. Williams and his men had won their first great victory.

So who was this man who dared to face down the Klan—and won? Six feet tall and 200 pounds, Williams was a military veteran, poet, radical activist, president of the local chapter of the NAACP, member of the NRA, and a college-educated member of the black middle class.

But more than anything else, Williams was a man who was not about to take the injustices of Jim Crow meekly. In an instantly infamous speech in 1959, Williams told a nervous nation just how he felt—and just how he planned to respond to lynchings and other racially motivated violence.

"We get no justice under the present system. If we feel that injustice is done, we must right then and there on the spot be prepared to inflict punishment on these people," he said after a white jury decided after a forty-five-minute deliberation to acquit a man who had attempted to rape a pregnant black woman in the presence of her five children. "Since the federal government will not bring a halt to lynching in the South and since the so-called courts lynch our people legally, if it's necessary to stop lynching with lynching, then we must be willing to resort to that method."

Williams's words set off an international firestorm, as newspapers around the world looked to the young leader of the Monroe NAACP as an omen of a new age of black militancy. And while his words were greeted with applause and support from many across the country—including Malcolm X—they also set Williams on a collision course with the mainstream civil rights movement, which was pushing nonviolence and attempting to win over white moderates.

In 1961, the two strategies came into sharp relief in a heated exchange between Williams and Martin Luther King Jr.

Seven blacks and six whites set out to ride buses from Washington, D.C., to New Orleans to test the Supreme Court's recent decision prohibiting segregation by interstate travel operators. As they stopped in Alabama for gas, a white mob attacked, slashing the tires on the bus. The bus driver attempted to drive away, but a pack of fifty cars operated by Klansmen pursued the bus to a stop. There, on the side of the road, a white mob beat the nonviolent protesters to a pulp, only stopping when the Alabama National Guard showed up and fired warning shots in the air.

Martin Luther King Jr., the national spokesman for nonviolence in the civil rights movement, flew down to speak. When he arrived, students begged him to join the rides, to risk personal injury to show the nation just how bad racial problems were in the American South. King declined.

And Williams spoke out. "No sincere leader asks his followers to make sacrifices that he himself would not make. You are a phony," he wrote in a telegram to King. "If you are the leader of this nonviolent movement, lead the way by example."

* * *

Today, students across the country learn to recite speeches by Martin Luther King Jr. They learn about nonviolence and peaceful resistance. And they learn how white people and black people joined hands to end Jim Crow and bring about a new era of racial equality. They don't learn about Robert Williams.

For while Williams's armed self-defense movement—and Black Power generally—contributed greatly toward the struggle for civil rights, he was also a radical who lived for many years in exile in Cuba. He was hunted down by the FBI, jailed, harassed, and cast out by his own country.

While King's approach toward civil rights embraced American ideals and appealed to the nation's best self, Williams's approach caustically pointed out the hypocrisies, evils, and injustices of the nation—often through alliances with America's enemies in the cold war.

After the Bay of Pigs invasion, the Cuban foreign minister

read a telegram from Williams addressed to the American delegate to the United Nations, Adlai Stevenson. "Now that the United States has proclaimed support for people willing to rebel against oppression, oppressed Negroes of the South urgently request tanks, artillery, bombs, money and the use of American airfields and white mercenaries to crush the racist tyrants who have betrayed the American Revolution and the Civil War. We also request prayers for this undertaking." Suffice it to say, America's political establishment did not look fondly on such interventions into cold war policy.

Yet while Williams and King promoted vastly different strategies, their goals were the same: equal rights and African American freedom. Even 100 years after the Emancipation Proclamation, the United States still sanctioned and promoted racial inequality, and turned a blind eye toward the ways in which violence was used to enforce that inequality. Between 1882 and 1965, white Southerners lynched close to 4,000 African Americans—and the American government did little, if anything, to prevent this violent enforcement of Jim Crow rule.

Robert F. Williams, like Kook and Quamana, like Charles Deslondes, took up arms against the United States of America in the name of freedom. They fought against U.S. government agents, they supported the overthrow of legally sanctioned racism, and they were exiled or executed for their actions.

Through slave revolts, armed political organizations like the Union League or Marcus Garvey's Universal African Legion, and through the Black Power movement of the post–World War II era, black Americans have had to fight every step of the way for their civil rights—from the right to eat in the same restaurant as whites to their right to not be sold into slavery.

Coming to terms with American history means addressing the 1811 uprising and the story of Robert F. Williams—not brushing these events under the rug because they upset safe understandings about who we are as a nation.

Their stories provide a different insight into history. These were men who stood for what they believed in and were willing to die for those beliefs. While today we see their beliefs as righteous, in their own time they were despised, exiled, even beheaded. These were people on the outside of traditional history—and the outside of the cultures they lived in. But their outsider status does not make them any less significant. Rather, their actions stand as a testament to the strength of the ideals of freedom and equality.

ACKNOWLEDGMENTS

This book began as my senior thesis, and my first debt is to the teachers and professors who guided me on my academic career. Sam Schaffer first introduced me to Southern history. Walter Johnson honed my analytical and theoretical skills. Vincent Brown helped me put the South in proper context. Tim McCarthy helped me tone down my polemical tendencies and find my voice. Drew Faust taught me everything I know about the Civil War—and a lot about the historian's craft. Dan Wewers applied a careful and skeptical eye to my writing. John Stauffer inspired me, encouraged me, and supported me at every turn, reading draft upon draft of first my thesis and then this book. I relied heavily on the trailblazing research of Gwendolyn Midlo Hall, Robert Paquette, Albert Thrasher, and Leon Waters. But my greatest thanks go to Susan O'Donovan. In no small part, she taught me how to think and write about slavery and American history. More than that, she has been a caring and dedicated mentor and friend.

My freshman year, I was lucky enough to take a seminar

from one of Harvard's most brilliant and devoted professors, Philip Fisher. He gave me crucial encouragement and advice, and he was the first to suggest that my thesis could become a book. I will always owe Joe Flood for seeing the seeds of this book in my thesis and for introducing me to my agent, Larry Weissman. Larry has been a committed advocate, a patient advisor, and a sage guide through a complex process. Larry connected me with Tim Duggan at Harper. Tim is a phenomenal editor, and he has helped this book realize its full potential. Allison Lorentzen and the rest of the team at Harper have been a pleasure to work with. Philip Hodges took an afternoon to help me with the author photo.

I would also like to thank my friends. Working with Diana Kimball in our Southern history colloquium with Professor O'Donovan was provocative, exciting, and above all fun. Vince Eckert, Mark Pacult, and Simon Williams all edited drafts of the book and gave me insightful criticism. Lewis Bollard and Sam Kenary have provided ample proof of Saint Paul's admonition to the Corinthians: as iron sharpens iron, so one man sharpens another. They both put up with three years of listening to me talk about this book, and their advice, support, and friendship have had a tremendous impact both on my writing and on me.

In addition, I want to thank Davis Kennedy, David Michaelis, Henry Louis Gates Jr., Brian Chingono, Glenda Gilmore, Charlie Young, Evan Thomas, Frank DeSimone, Emmet McDermott, Eric Foner, Nicki Bass, Adam Rothman, Amar Bakshi, Rebecca Scott, Mac Bartels, Peter Trombetta, Jeffrey Thornton, and Balmore Toro, all of whom helped shape this book in one way or another.

Finally, I would like to thank my family. Uncle Bill and Aunt Isabelle hosted me at their home in Little Compton when I was writing my thesis. I wrote this book with Rob, Lisa, and Tom in mind: if they enjoy reading this book, I know I will have succeeded. Willy has been an inspiration to me since I was little. He encouraged me to take up journalism, coached my writing, advised me at Harvard, and has edited probably five drafts of this book. Dad, similarly, has been a constant editor and advisor and I could not have done this without his help and formidable insight. I will always be grateful to my mom for sharing her love of reading with me, and, more important, for always believing in me. I hope that this book reflects well the contributions of all of these brilliant and generous professors, friends, and family members.

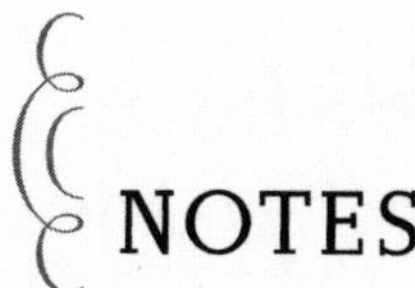

NOTES

PROLOGUE

1 "Though the cause . . . is strong": Lowell, *Poetical Works*, 68.

2 a mere twenty-four pages: Paquette, "The Great Louisiana Slave Revolt."

1. CARNIVAL IN NEW ORLEANS

9 "The river . . . for empire": Barry, *Rising Tide*, 97.

10 Down from the mountains . . . to New Orleans: Barry, *Rising Tide*, 38–39.

10 prime entrepot: Meinig, *The Shaping of America*, 15–16.

10 "the produce . . . to market": Greeley, *American Conflict*, 55.

10 Magnolias, orange trees . . . columned porticoes: Crété, *Daily Life in Louisiana*, 253.

11 "The social status . . . 'good natured'": de Laussat, *Memoirs of My Life*, 61.

12 The roads . . . games in the field: Crété, *Daily Life in Louisiana*, 97.

12 "most active . . . the country": de Laussat, *Memoirs of My Life*, 53–54.

12 small talk and gossip: de Laussat, *Memoirs of My Life*, 20.

12 unparalleled in the United States: Sitterson, *Sugar Country*, 45.

13 Every year . . . alcohol: Crété, *Daily Life in Louisiana*, 204–5.

13 "You never saw anything more brilliant": de Laussat, *Memoirs of My Life*, 86.

13 Slaves brought in . . . the next morning: de Laussat, *Memoirs of My Life*, 81.

13 almost entirely to dancing and gambling: Kinser, *Carnival, American Style*, 22.

14 As early as the 1740s . . . to dance: Kinser, *Carnival, American Style*, 17, 59.

14 Jean Noël Destrehan's relatives . . . as a sugar planter: Harvey, Roger, and D'Oliveira, *To Reach Afar*, 2–3, 32.

15 "Those who have . . . annual expences": Sitterson, *Sugar Country*, 158.

16 "We could not imagine . . . opulence and luxury": Gayarré, *History of Louisiana*, 4:59.

17 Awakening at sunrise . . . refresh him: Stedman, *Stedman's Surinam*, 104.

17 "was there . . . in the back": de Laussat, *Memoirs of My Life*, 61.

17 "cultivation must cease . . . demolished habitations": Gayarré, *History of Louisiana*, 4:62.

17 75 percent: Yoes, *Louisiana's German Coast*, 72.

17 close to 90 percent: Rothman, *Slave Country*, 108; Conrad, *German Coast*, 108.

18 "To the necessity . . . to whites": Gayarré, *History of Louisiana*, 4:62.

2. PATHS TO SLAVERY

19 African and African-descended slaves . . . along the riverbank: Latrobe, *Impressions Respecting New Orleans*, 21–22.

20 The participants . . . the drummers: Latrobe, *Impressions Respecting New Orleans*, 50.

20 rocked the Crescent City: Sublette, *The World That Made New Orleans*, 3.

21 The men . . . the celebrants: Kinser, *Carnival, American Style*, 36, 41.

21 leaders had been chiefs or kings in Africa: Childs, *1812 Aponte Rebellion in Cuba*, 117.

21 "He wags his head . . . upon the multitude": Kinser, *Carnival, American Style*, 35.

22 Their names . . . with the ocean: Summer Institute of Linguistics Aukan–English Dictionary; Bartle, "Forty Days."

22 The Asante kingdom controlled . . . from birth": Fynn, *Asante and Its Neighbours,* 1–4, 32–33.

22 born around 1790: American Uprising Slave Database. Conrad, *German Coast;* Thrasher, *On to New Orleans!*

23 In the Lower Guinea . . . King Jose I: Thornton, "African Soldiers in the Haitian Revolution."

23 ten separate slave ships . . . from Charleston, South Carolina: Leglaunec, "Slave Migrations."

23 From about 1770 . . . the Windward Coast: Leglaunec, "Slave Migrations."

24 Some slaves might have . . . baptized elsewhere: Leglaunec, "Slave Migrations."

24 Forty percent . . . fourth year of labor: Miller, *Way of Death,* 440–41.

25 records suggest he was actually born in South Carolina: Carretta, *Equiano, the African.*

25 By his telling, Equiano . . . animal skins: Equiano/Sollors, *Gustavus Vassa,* 20–23.

26 "We are almost" . . . golden jewelry as well: Equiano/Sollors, *Gustavus Vassa,* 22.

26 Every man, woman, and child . . . their plantings and harvestings: Equiano/Sollors, *Gustavus Vassa,* 26.

27 One day . . . "with our tears": Equiano/Sollors, *Gustavus Vassa,* 32.

27 "My surprise" . . . terrified: Equiano/Sollors, *Gustavus Vassa,* 39–41.

28 "I was now . . . loose hair": Equiano/Sollors, *Gustavus Vassa,* 39–41.

28 Some slaves choked . . . and died: Harms, *The Diligent,* 252.

28 The slave traders brought . . . new oppressions: Harms, *The Diligent,* 253.

29 The captain . . . ran high: Harms, *The Diligent,* 261, 267.

29 The crew . . . their destination: Harms, *The Diligent,* 268.

30 "to teach a lesson to all the others": Harms, *The Diligent,* 270.

30 "It was usual . . . among one another": Harms, *The Diligent,* 297–98.

31 Prior to their arrival . . . before sale: Kiple, *The Caribbean Slave*, 57.

32 William Kenner . . . merchant firm: Kane, *Plantation Parade*, 25–26.

32 a full-service business: Smith and Smith, *Cane, Cotton & Crevasses*, 26.

32 Brown had . . . for sugar production: Sitterson, *Sugar Country*, 23.

32 "towering . . . exceedingly unpopular": "Memoirs of Micah Taul," in W. C. C. Claiborne, *Interim Appointment*, 264.

33 The slaves tried to imagine . . . walking sticks: W. Johnson, *Soul by Soul*, 165–66.

34 Joseph the Spaniard: St. Charles Parish, *Original Acts*, hereafter cited as Denunciations.

34 "While furnishing . . . have resulted": Sublette, *The World That Made New Orleans*, 88.

35 "Nothing is more dreaded . . . plot their rebellions": Sublette, *The World That Made New Orleans*, 73.

35 In fact . . . declaring war: Thornton, "African Dimensions of the Stono Rebellion," 1112.

35 In 1812 . . . "but dead": Childs, *1812 Aponte Rebellion in Cuba*, 123.

36 "headmen" . . . "primal powers of thunder and lightening": Linebaugh and Rediker, *The Many-Headed Hydra*, 184–85.

3. A REVOLUTIONARY FORGE

39 123 million pounds . . . and rum: James, *The Black Jacobins*, 45.

40 The island accounted for . . . refined goods: James, *The Black Jacobins*, 48–49.

41 "How can we make . . . men and animals": Dubois, *Avengers of the New World*, 93.

42 "One has to hear . . . surrounds him": James, *The Black Jacobins*, 18.

42 "*Eh! Eh!* . . . this vow": James, *The Black Jacobins*, 18.

42 On the night . . . "spread like a torrent": Dubois, *Avengers of the New World*, 94.

43 "strange eyes . . . hearts of all of us": Dubois, *Avengers of the New World*, 100.

43 In the first eight . . . 80,000: Dubois, *Avengers of the New World*, 113.

44 "There is a motor . . . come to know": Pierre Mossut to Marquis de Gallifet, September 19, 1791 (Archives Nationales, 107 AP 128), in Dubois and Garrigus, *Slave Revolution in the Caribbean*, 94.

44 In control of France . . . 42,000 battle-hardened men: Dubois, *Avengers of the New World*, 251, 260.

45 "war of extermination": Dubois, *Avengers of the New World*, 290.

45 "ignorant of color prejudice": Dubois, *Avengers of the New World*, 292–93.

45 Over 80 percent: Paul Lachance, "Repercussions of the Haitian Revolution in Louisiana," in Geggus, *Impact of the Haitian Revolution*, 210.

46 "Let us imitate . . . territory of liberty": Jean-Jacques Dessalines, "The Haitian Declaration of Independence," in Dubois and Garrigus, *Slave Revolution in the Caribbean*, 188–91.

46 In 1789 . . . only 9,000: Paul Lachance, "Repercussions of the Haitian Revolution in Louisiana," in Geggus, *Impact of the Haitian Revolution*, 211.

46 "Damn sugar, damn coffee, damn colonies": Robert Paquette, "Revolutionary Saint Domingue in the Making of Territorial Louisiana," in Gaspar and Geggus, *A Turbulent Time*, 209.

47 41 percent of the North American continent: Barry, *Rising Tide*, 21.

47 By 1802 . . . sugar per year: Paul Lachance, "Repercussions of the Haitian Revolution in Louisiana," in Geggus, *Impact of the Haitian Revolution*, 211.

48 As Americans and Europeans . . . raw sugar: Mintz, *Sweetness and Power*, 120.

48 In a few short years . . . revolved around sugar: Berlin, *Many Thousands Gone*, 199, 325.

4. EMPIRE'S EMISSARY

52 "We have lived long . . . first rank": Geer, *The Louisiana Purchase*, 197.

53 "puny force . . . subject for ridicule": Young, "The United States Army in the South," 99, 105.

53 "The prejudices . . . American": *Orleans Gazette for the Country*, June 6, 1811.

53 devoted to a single, slave-made staple crop: Berlin, *Many Thousands Gone*, 325.

53 "guide the rising generation . . . virtue": Hatfield, *William Claiborne*, 112.

54 "glowing colours": N. H. Claiborne, *Notes on the War*, 92.

54 "do nothings" . . . "American liberty": Hatfield, *William Claiborne*, 4.

54 "Dear my country . . . my country": N. H. Claiborne, *Notes on the War*, 93.

54 imperial colony of alien people: Meinig, *The Shaping of America*, 15.

55. "The very trees . . . first proprietors": N. H. Claiborne, *Notes on the War*, 102.

55 "This would not sweeten . . . as well as theirs": Meinig, *The Shaping of America*, 15.

55 "All Louisianians are Frenchmen at heart": de Laussat, *Memoirs of My Life*, 18.

55 *chacas, catchoupines, catchumas,* and *kaintucks*: Crété, *Daily Life in Louisiana*, 70.

55 When Claiborne arrived . . . meant to be American: Hatfield, *William Claiborne*, 114.

56 "mischiefs . . . every other passion": Hatfield, *William Claiborne*, 120.

56 "Hail Columbia" . . . "attended my administration": Kinser, *Carnival, American Style*, 29.

57 "the insignificant part he acted in the circle": Hatfield, *William Claiborne*, 153–154.

57 "oppressive and degrading" . . . "enjoy it with safety": Gayarré, *History of Louisiana*, 4:59.

57 "who neither" . . . "could personally communicate": Gayarré, *History of Louisiana*, 4:63.

58 "to whose conversation . . . devoted": N. H. Claiborne, *Notes on the War*, 97.

58 "We were among . . . composed it": Gayarré, *History of Louisiana*, 4:60.

58 "Annexed to your country . . . allegiance of citizens": Gayarré, *History of Louisiana*, 4:64.

58 "The people . . . better ascertained": Gayarré, *History of Louisiana*, 4:67.

59 "a hazardous experiment": Gayarré, *History of Louisiana*, 4:68.

59 "the lowest Indian tribes . . . lower Louisiana": Harvey, Roger, and D'Oliveira, *To Reach Afar*, 42.

59 "Renegadoes . . . subject of accusation": N. H. Claiborne, *Notes on the War*, 110.

60 "pecuniary difficulty" . . . "graves of Louisiana": Hatfield, *William Claiborne*, 156–57.

5. CONQUERING THE FRONTIER

61 "We should have . . . self government": Jefferson, *Life and Writings*, 185.

62 "intruder king": Weber, *The Spanish Frontier in North America*, 296.

64 "It would be . . . *such a request?*": W. C. C. Claiborne, *Official Letter Books*, 31.

64 "friendly disposition . . . decided measures": W. C. C. Claiborne, *Official Letter Books*, 32.

64 "lose no time . . . right direction": W. C. C. Claiborne, *Official Letter Books*, 32.

64 "Endowed . . . bone and muscle": Kennedy, *Mr. Jefferson's Lost Cause*, 218.

65 "more satisfactory" . . . "to the palate of the administration": Kennedy, *Mr. Jefferson's Lost Cause*, 224.

65 Before dawn . . . Louis de Grand Pré: Arthur, *The Story of the West Florida Rebellion*, 103.

66 "Hurrah Washington!": Arthur, *The Story of the West Florida Rebellion*, 105–6.

66 "Betrayed . . . independent State": Kennedy, *Mr. Jefferson's Lost Cause*, 225.

67 "the intrigues . . . and fortunes": W. C. C. Claiborne, *Official Letter Books*, 35.

67 "protect them . . . and religion": W. C. C. Claiborne, *Official Letter Books*, 49.

68 "the association of the Individuals" . . . "towards Spain": W. C. C. Claiborne, *Official Letter Books*, 61.

69 "absolute anarchy": W. C. C. Claiborne, *Official Letter Books*, 85.

69 A gentleman . . . the territory: W. C. C. Claiborne, *Official Letter Books*, 91.

69 Reports . . . of Carnival: W. C. C. Claiborne, *Official Letter Books*, 88.

6. MASKS AND MOTIVES

71 "For it is the same... and advise them": Robert Paquette, "The Drivers Shall Lead Them: Image and Reality in Slave Resistance," in Paquette and Ferleger, *Slavery, Secession, and Southern History*, 31.

71 set back a short distance ... summer sun: Bernhard, *Travels Through North America*, 2:80.

72 thirty to forty yards... travelers and traders: Brackenridge, *Views of Louisiana*, 176–77.

72 Docks ... transportation systems: Hall, *Africans in Colonial Louisiana*, 120.

72 Property lines... swamps: Hall, *Africans in Colonial Louisiana*, 120.

72 Irrigation ditches... rectangles: Rehder, *Delta Sugar*, 158.

72 The Mississippi... vegetable mold: Sitterson, *Sugar Country*, 14.

73 As was standard... survive and work: Kane, *Plantation Parade*, 160–61.

73 "whether actually in the field... on the watch": Northup/Eakin and Logsdon, *Twelve Years a Slave*, 172.

73 By keeping constant watch ... over their actions: "Thanks to the techniques of surveillance, the 'physics' of power, the hold over the body, operate according to the laws of optics and mechanics, according to a whole play of spaces, lines, screens, beams, degrees and without recourse, in principle at least, to excess, force or violence." Foucault, *Discipline and Punish*, 177.

74 "It is fair to say ... insolence of their authority": Quoted in Robert Paquette, "The Drivers Shall Lead Them: Image and Reality in Slave Resistance," in Paquette and Ferleger, *Slavery, Secession, and Southern History*, 33.

74 "human bloodhounds": Robert Paquette, "The Drivers Shall Lead Them: Image and Reality in Slave Resistance," in Paquette and Ferleger, *Slavery, Secession, and Southern History*, 37.

75 industrial expertise on a level with that of Northern factories: Mintz, *Sweetness and Power*, 47.

76 sixteen or more hours per day, seven days a week: McDonald, *Economy and Material Culture of Slaves*, 14–15.

76 sweet, juicy stalks... fifteen feet: Mintz, *Sweetness and Power*, 21.

76 Working round the clock . . . into January: Moody, *Slavery on Louisiana Sugar Plantations*, 48–49.

77 "by a wise distribution . . . any of them": Colonial prefect Clement de Laussat, quoted in Follett, *The Sugar Masters*, 18.

77 tropical disease: Sitterson, *Sugar Country*, 92.

77 "From June to the beginning of October . . . in the country": Latrobe, *Impressions Respecting New Orleans*, 141–42.

78 "More than once . . . voice against it": Northup, *Twelve Years a Slave*, 104–5.

78 The complexity . . . management styles: Follett, *The Sugar Masters*, 92.

78 "The feelings of humanity . . . whips in hand": de Laussat, *Memoirs of My Life*, 54.

78 A first punishment . . . recalcitrant slaves": Sitterson, *Sugar Country*, 89.

79 "Three stakes . . . every stroke": Rothman, *Slave Country*, 95–96.

80 Sugar work was too grueling . . . natural reproduction: "A perception prevailed, true or not, that it was cheaper to work field slaves to death in five years or so and replace them by purchase than to see to their long-term maintenance and reproduction," wrote historian Robert Paquette about Cuban sugar plantations. Paquette, *Sugar Is Made with Blood*, 55.

80 In 1800 . . . less important: Sitterson, *Sugar Country*, 158.

80 first on the Atlantic slave trade and then on the internal slave trade: An illegal Atlantic slave trade did continue. "Like elsewhere throughout South and Central America and the Caribbean, in Jamaica slaves put to cultivating sugar died faster than they bore progeny," wrote historian Roderick McDonald. "Only the slave trade, the Black Mother, could maintain and increase the size of slave populations. While Jamaica relied on slave traffic across the Atlantic, for Louisiana the Black Mother was interstate traffic, the slave states of the Old South supplying the men, women and children the sugar planters needed." McDonald, *Economy and Material Culture of Slaves*, 15.

80 Those who complied . . . extra food: Robert Paquette, "The Drivers Shall Lead Them: Image and Reality in Slave Resistance," in Paquette and Ferleger, *Slavery, Secession, and Southern History*, 33.

7. THE REBELS' PACT

84 take up with a woman: Kaye, *Joining Places*, 60.

84 he was the son of a white planter: Unlike most slaves, who bore only a first name, Charles used his deceased master's name, Deslondes. This likely indicates that Deslondes was his father.

84 "one great brothel": W. Phillips, *Speeches, Lectures, and Letters*, 108.

86 As the white planters . . . on the German Coast: Conrad, *The German Coast*, 108.

86 to kill all the whites: Conrad, *The German Coast*, 106.

87 Posting a spy . . . organize the uprising: Thrasher, *On to New Orleans!*, 3.

87 A twenty-five-year-old carpenter . . . closest to New Orleans: Harry was a "rough carpenter" who was "well-acquainted with the business of a sugar plantation." "Sound and healthy" at the age of twenty-five, he was valued at $800. He was executed after being tried at the St. Charles Parish Tribunal. American Uprising Slave Database.

87 During their free time . . . in the marketplaces: McDonald, *Economy and Material Culture of Slaves*, 69.

89 "Woe to the white" . . . for the German Coast slaves: Hall, *Africans in Colonial Louisiana*, 213, 220, 230, 232.

89 In 1795, the Spanish discovered . . . in the marketplaces: Hall, *Africans in Colonial Louisiana*, 358.

89 "liberty, property, security, and resistance to oppression": Declaration of the Rights of Man.

89 The plot was discovered . . . Puerto Rico, and Cuba: Hall, *Africans in Colonial Louisiana*, 344.

90 The residents . . . relief to the planters: Gayarré, *History of Louisiana*, 4:118.

90 20,000 African slaves: Berlin, *Many Thousands Gone*, 344.

90 Kongo had been . . . warrior knowledge: Thornton, "African Dimensions of the Stono Rebellion," 1108–9.

90 The Kongolese had developed . . . their abilities: Dubois, *Avengers of the New World*, 109.

90 They used flags . . . the field of battle: Thornton, "African Dimensions of the Stono Rebellion," 1111.

8. REVOLT

97 a powerful rainstorm: "The wind being from the Northward and westward blowing at the same time fresh with considerable rain would have been directly ahead for vessels attempting to ascend the river." John Shaw to Paul Hamilton, New Orleans, 18 January 1811.

97 "half leg deep in Mud": Wade Hampton to the Secretary of War, New Orleans, January 16, 1811, in Thrasher, *On to New Orleans!*, 269.

97 Rain meant . . . from the swamps: Sitterson, *Sugar Country*, 19.

98 "This is how . . . stomachs of the whites": Childs, *1812 Aponte Rebellion in Cuba*, 117.

99 A high roof . . . gallery from the rain: Bernhard, *Travels Through North America*, 2:80.

99 three long cuts: Bernhard, *Travels Through North America*, 2:81.

100 the stores in the basement of Andry's mansion: The *Louisiana Gazette* reported that the slaves seized "the public arms that was in one of Mr. Andry's stores." *Louisiana Gazette* (New Orleans), January 17, 1811.

101 "On to New Orleans!": Thrasher, *On to New Orleans!*, 51.

102 Live oaks . . . marshy cypress swamps: Bernhard, *Travels Through North America*, 2:54.

103 knee-deep in mud: Wade Hampton to the Secretary of War, New Orleans, January 16, 1811, in Thrasher, *On to New Orleans!*, 269.

103 A group of ten slaves: One of these slaves, Theodore, was tried in the court at New Orleans, where he received mercy for "having made important discoveries, touching the late insurrection." American Uprising Slave Database; Yoes, *Louisiana's German Coast*, 67.

103 Waving his saber in the air: Denunciations.

103 Achille Trouard . . . to hide: Perret to Fontaine, January 13, 1811, *Moniteur de la Louisiane* (New Orleans), January 17, 1811, translated by Robert Paquette and Seymour Drescher, in Engerman, Drescher, and Paquette, *Slavery*, 324–26.

104 The slave Pierre . . . his life: Labranche recognized the dual nature of these slaves' presence in the swamps, adding a note of distrust to his reporting of his slave driver Pierre's sources. "These slaves having fled into the swamp back of the Labranche place to save themselves from the rebels, or so they told Pierre," he emphasized. Labranche

was well aware that this was an information network not entirely under his control, and that the information he received through it must be necessarily understood as colored by the intentions and motives of those creating it. Conrad, *The German Coast*, 107.

104 "flee immediately . . . [the] farm": Conrad, *The German Coast*, 107.

104 All knew . . . that much clear: Freehling, *The Road to Disunion*, 79.

105 In a panic . . .blazed by escaping slaves: Conrad, *The German Coast*, 107.

105 "torrent of rain and the frigid cold": Perret, in Engerman, Drescher, and Paquette, *Slavery*, 324.

105 "Freedom or death": Dubois, *Avengers of the New World*, 116.

106 half of James Brown's slaves: American Uprising Slave Database.

106 the average height . . . inches: Kiple, *The Caribbean Slave*, 58.

107 "The Assianthes . . . cold blood": Fynn, *Asante and Its Neighbours*, 143.

107 Drawn in small groups . . . Anglo-Americans: American Uprising Slave Database; Denunciations.

108 "there was a large number . . . killing whites": Labranche testified that Dominique told slaves along the way to warn their masters of the uprising. "Labranche added that he knew that while Dominique, Bernard Bernoudy's slave, was on his way home to alert his master, he stopped at Pierre Pain's farm and instructed Pain's slave Denys to warn as many whites as possible of the impending danger," read the court testimony. Conrad, *The German Coast*, 107.

108 When Dominique arrived . . . along the way: Conrad, *The German Coast*, 107.

108 Local legend . . . a black child: Kane, *Plantation Parade*, 128.

108 Trépagnier did not think . . . for the slaves to arrive: Charles Gayarré, *History of Louisiana*, 4: 267–68.

109 Kook led a party . . . to pieces: Denunciations; American Uprising Slave Database.

9. A CITY IN CHAOS

115 In the dense neighborhood . . . plantations of the German Coast: Bernhard, *Travels Through North America*, 2:54–55, 71.

116 Slaves formed the great . . . one side to the other: Bernhard, *Travels Through North America*, 2:56.

117 They had heard the stories . . . children alike: Langley, *The Americas in the Age of Revolution*, 112–13.

117 "miniature representation of the horrors of St. Domingo": *Baltimore American and Commercial Party Advertiser*, February 20, 1811.

117 "Sometime before noon . . . West Florida": Claiborne to General Wade Hampton, New Orleans, January 7, 1811, in W. C. C. Claiborne, *Official Letter Books*, 92.

117 He feared . . . slaves and free black people: "Message from the Mayor, January 12, 1811," in Thrasher, *On to New Orleans!*, 274.

118 "Sir . . . repass the same": Claiborne to General Wade Hampton, New Orleans, January 9, 1811, in W. C. C. Claiborne, *Official Letter Books*, 93.

118 "All the Cabarets . . . immediately closed": Claiborne, "General Orders," January 9, 1811, in Thrasher, *On to New Orleans!*, 267.

118 "People of color . . . under mask": Kinser, *Carnival, American Style*, 25.

118 "No male Negro . . . 6 o'clock": Claiborne, "General Orders," January 9, 1811, in Thrasher, *On to New Orleans!*, 267.

119 "about 12 O'Clock . . . beyond description": Hampton, in Thrasher, *On to New Orleans!*, 269

119 of equal ferocity to the revolutionaries of Haiti: François-Xavier Martin described "carriages, wagons and carts, filled with women and children . . . bringing the most terrible accounts." It was, he wrote, a "miniature representation of the horrors of St. Domingo." *Baltimore American and Commercial Party Advertiser*, February 20, 1811.

119 "weak detachment . . . by the Rioters": John Shaw to Paul Hamilton, New Orleans, 18 January 1811.

120 "All were on the alert . . . and property": John Shaw to Paul Hamilton, New Orleans, 18 January 1811.

120 "I pray God . . . murdering career": Claiborne to Major St. Amand, New Orleans, January 9, 1811, in W. C. C. Claiborne, *Official Letter Books*, 93–94.

121 a party of volunteer cavalry: New Orleans City Court, Case No. 195, February 18, 1811, in Thrasher, *On to New Orleans!*, 246–47.

122 Riding along the River Road . . . chaos of the German Coast: New Orleans City Court, Case No. 195, February 18, 1811, in Thrasher, *On to New Orleans!*, 246.

10. A SECOND WIND

124 Augustin, a highly valued sugar worker: He was valued at $1,000. American Uprising Slave Database.

124 Horses were powerful military tools: For more on the military advantages of horses, see Law, "Horses, Firearms, and Political Power."

124 At the plantation of Butler and McCutcheon . . . with his family: The rebel Simon had escaped from the plantation before. Simon was "lately from Baltimore, about 20 years of age, 5 feet 6 or 7 inches high, has a scar on his left cheek, and one on his forehead, handsome features." Brought to New Orleans by the internal slave trade, Simon had tried to escape back to his birthplace and presumably his family. *Louisiana Gazette* (New Orleans), July 24, 1810, in Thrasher, *On to New Orleans!*, 166.

124 Dawson . . . Joe Wilkes: American Uprising Slave Database.

125 Jasmin, Chelemagne, and Gros and Petit Lindor: American Uprising Slave Database.

125 Rubin and Coffy: American Uprising Slave Database.

125 "to keep an eye on the situation": Conrad, *The German Coast*, 107.

126 threatening to kill any slaves that would not join: Denunciations.

126 The rebels knew . . . violence, too: Dagobert, a slave owned by Delhomme, testified that "except for the ones whom he denounced for having marched of their own free will, he believes that the others whom he accused were forced to march." In the trial of a runaway a month after the uprising, Étienne Trépagnier's slave Augustin "stated that he had nothing to do with the recent insurrection; that during the event he was taken by some blacks who threatened him and demanded to know the name of his master." Denunciations; Conrad, *The German Coast*, 108.

126 set fire to the home of the local doctor: Denunciations; American Uprising Slave Database.

126 very different approaches to medicine and healing: "Slaves were commonly used as medical doctors and surgeons in eighteenth-

century Louisiana," wrote Gwendolyn Midlo Hall. "They were skilled in herbal medicine and were often better therapists than the French doctors, who were always described as surgeons." Hall, *Africans in Colonial Louisiana*, 126.

126 But the slaves . . . white medicine: Alexandre Labranche wrote that "he lost a house which was occupied by the doctor, located near Pierre Reine's line; burned by the brigands, valued at $1,000." Conrad, *The German Coast*, 109–10.

126 wealthiest and largest plantation: "When Louis-Augustin Meuillion, probably the largest slaveholder on the Coast, died in 1811, his succession inventory listed fewer than one hundred slaves," wrote Conrad. Conrad, *The German Coast*, viii.

126 pillaging and destroying: "The sale of household objects did not conform to the inventory because, during the slave uprising of January 9, the house was entered and pillaged." Conrad, *The German Coast*, 102.

127 "did alone . . . of the late Meuillion": Conrad, *The German Coast*, 104.

127 Half Native American: Bazile is described as a *griffe*, or a black-Indian mixture. American Uprising Slave Database; Hall, *Africans in Colonial Louisiana*, 113.

127 Cannes Brûlées: "With one Benjamin Morgan, [William Kenner] acquired land a few miles upriver from New Orleans in an area called 'Cannes-Brulees' (Land of the Burnt Canes) so-named from the Indians' historic practice of torching marsh-grass canes to flush out their game." Smith and Smith, *Cane, Cotton & Crevasses*, 25.

127 "most outstanding brigands": Denunciations.

127 Harry garnered . . . hand, Harry: American Uprising Slave Database; Denunciations.

127 all the black males: Denunciations.

127 Lindor, a coachman and carter: American Uprising Slave Database.

128 124 individual slaves: The survey of planters conducted by the St. Charles Parish planters indicates that 124 slaves were involved in the 1811 insurrection. Eyewitnesses reported up to 500. American Uprising Slave Database.

128 rivaling the size of the American military force in the region: Young, "The United States Army in the South."

128 been employed as unskilled or low-skilled workers: In terms of occupation, there was a fair mix. Field hands, cartmen, and sugar workers dominated the roster, with many other occupations appearing occasionally. Field hands, cartmen, sugar workers of various types, plowmen, and shovel, pickaxe, and axe workers seem to have been the most common occupations. American Uprising Slave Database.

128 "only one half . . . cane knives": *Louisiana Gazette* (New Orleans), January 17, 1811.

129 "The Brigands . . . Sugar works": Wade Hampton to the Secretary of War, New Orleans, January 16, 1811, in Thrasher, *On to New Orleans!*, 269–70.

129 "a few young men . . . great silence": Hampton to the Secretary of War, in Thrasher, *On to New Orleans!*, 269–70.

130 "killing poultry . . . and rioting": *Richmond Enquirer*, February 22, 1811. This article is a reproduction of a piece in the *Louisiana Gazette*, the original of which is practically unreadable.

130 Warfare . . . better-armed forces: Dubois, *Avengers of the New World*, 108.

130 pursue the fugitives: *Richmond Enquirer*, February 22, 1811.

11. THE BATTLE

135 "My poor son . . . of that nature": Misspellings are original to the document. Manuel Andry to Claiborne, New Orleans, January 11, 1811, in Thrasher, *On to New Orleans!*, 268.

136 "halt the progress of the revolt": Perret, in Engerman, Drescher, and Paquette, *Slavery*, 324–26.

137 "forced march": "About 9 o'clock of the same Morning they were fallen in with by a spirited party of Young Men from the opposite side of the river, who fired upon & disperse them, Killing some 15, or 20, & wounding a great many more," wrote Hampton. Wade Hampton to the Secretary of War, New Orleans, January 16, 1811, in Thrasher, *On to New Orleans!*, 269–70.

137 "Let those who are willing . . . move out!": Perret to Fontaine, in Engerman, Drescher, and Paquette, *Slavery*, 324–26.

139 "In action . . . three hundred feet distance": Lewis, *Small Arms and Ammunition*, 305.

138 "We are now fighting . . . die first": Redkey, *A Grand Army of Black Men*, 147–48.

138 "The blacks . . . in line": Report of Spanish Consul, January 13, 1811.

138 *Recover arms . . . aim, fire*: Lewis, *Small Arms and Ammunition*, 301.

139. clouds of smoke: Hess, *The Union Soldier in Battle*, 8.

139 Guns roared . . .invisible: Hess, *The Union Soldier in Battle*, 130–31.

139. Their hair . . . their faces: Hess, *The Union Soldier in Battle*, 8.

140 "Fifteen or twenty . . . into the woods": Report of Spanish Consul, January 13, 1811, Eusebio Bardari y Azara to Vicente Folch, February 6, 1811.

140 "left 40 to 45 men . . . several chiefs": Perret to Fontaine, in Engerman, Drescher, and Paquette, *Slavery*, 324–26.

140 "considerable slaughter": Andry to Claiborne, in Thrasher, *On to New Orleans!*, 268.

140 "I was desolate . . . prospect before me": Northup/Eakin and Logsdon, *Twelve Years a Slave*, 100.

141 party of Native Americans: *Raleigh Star*, February 24, 1811. This was a common strategy in Louisiana's maroon wars. Hall, *Africans in Colonial Louisiana*, 365–66. The Indians often sided with the Europeans in wars against the slaves, because their own ideology portrayed slaves as outcasts deserving of little sympathy. "Natchez Indians had their own notions of slavery, as did the neighboring Choctaw. American Indian forms of slavery were different from those employed by Europeans in the Americas. The Natchez and Choctaw viewed slavery in terms of membership in (or exclusion from) society." Libby, *Slavery and Frontier Mississippi*, xii.

141 "I left . . . who had fled": Perret to Fontaine, in Engerman, Drescher, and Paquette, *Slavery*, 324–26.

141 "I never knew . . . by the dogs": Northup/Eakin and Logsdon, *Twelve Years a Slave*, 101.

141 They discovered . . . cold and terror: Perret to Fontaine, in Engerman, Drescher, and Paquette, *Slavery*, 324–26.

142 "long, savage yells . . . sinking into my flesh": Northup/Eakin and Logsdon, *Twelve Years a Slave*, 101–2.

142 "the principal leader of the bandits": Denunciations.

142 According to one witness . . . pile of straw": Samuel Hambleton to David Porter, January 15, 1811. Papers of David Porter, Library of Congress, in Engerman, Drescher, and Paquette, *Slavery*, 326.

142 "Pierre Griffe" . . . to the Andry estate: Perret to Fontaine, in Engerman, Drescher, and Paquette, *Slavery*, 324–26.

143. In the days . . . Spanish in West Florida: Kastor, *The Nation's Crucible*, 102

143 Milton had heard . . . to the militia: Wade Hampton to William Claiborne, January 12, 1811, in Thrasher, *On to New Orleans!*, 269.

143 "I have Judged . . . higher Up": Hampton to Claiborne, January 11, 1811, in Thrasher, *On to New Orleans!*, 269.

143 "The [slaves'] plan . . . more formidable": Hampton to Claiborne, New Orleans, January 12, 1811, in Thrasher, *On to New Orleans!*, 269.

144 "proprietors . . . maintain order": Perret to Fontaine, in Engerman, Drescher, and Paquette, *Slavery*, 324–26.

12. HEADS ON POLES

147 "There it was . . .on their sticks": Conrad, *Heart of Darkness and The Secret Sharer*, 132–33.

148 "They were brung . . . long poles": Hambleton to Porter, in Engerman, Drescher, and Paquette, *Slavery*, 326.

148 Those passersby . . . state in the making: In the words of Katherine Verdery, bodies have an "ineluctable self-referentiality as symbols: because all people have bodies, any manipulation of a corpse directly enables one's identification with it through one's own body, thereby tapping into one's reservoirs of feeling." Verdery, *The Political Lives of Dead Bodies*, 32–33.

148 "Had not the most prompt . . . waste by the Rioters": John Shaw to Paul Hamilton, New Orleans, 18 January 1811.

149 "make a GREAT EXAMPLE": Andry to Claiborne, January 11, 1811, in Thrasher, *On to New Orleans!*, 268.

149 This was not just a French . . . and to Africa: In New England, colonists and Indians communicated with each other through corpses. "When English soldiers came upon English heads on poles, they often simply took them down and put Indian heads in their place,"

wrote historian Jill Lepore. Lepore, *The Name of War*, 180. In Jamaica and the other British sugar islands of the Caribbean, power was physically manifested through beheadings. "The frequency of mutilations and aggravated death sentences, which in eighteenth-century England were reserved for traitors, signaled the expansion of the very concept of treason to include almost any crime committed by slaves," wrote Vincent Brown. V. Brown, *The Reaper's Garden*, 140. In the African kingdom of Dahomey, where many slaves came from, kings accumulated the skulls of defeated enemies and used them as architectural decorations. Law, "'My Head Belongs to the King'."

150 From 1760 . . . Danish territories: Linebaugh and Rediker, *The Many-Headed Hydra*, 193.

150 The Coromantee slave . . . on a pole: Linebaugh and Rediker, *The Many-Headed Hydra*, 222.

150 When slaves rebelled . . . heads on stakes: Dubois, *Avengers of the New World*, 116.

150 In 1795 . . . Pointe Coupée: Hall, *Africans in Colonial Louisiana*, 344.

151 "Condemnation . . . again established": John Shaw to Paul Hamilton, New Orleans, 18 January 1811.

151 Witnesses to these spectacles . . . public decay: "The spectacular violence of slavery was both a political and aesthetic discourse which was grounded in eighteenth-century notions of a triangular violent gaze: most bloody vignettes utilized a visual and moral interplay between victim, perpetrator and spectator," wrote literary historian Ian Haywood. "Spectacular violence existed uneasily but powerfully on the borders between reality and fantasy, reportage and representation, aesthetic gratification and political mobilization." Haywood, *Bloody Romanticism*, 58.

152 "awaiting the stroke of law . . . promptly destroyed": Denunciations.

152 a tribunal of slaveholders: John Destrehan, Alexandre Labranche, Pierre-Marie Cabaret de Trepy, Adelard Fortier, and Edmond Fortier joined St. Martin in conducting the tribunal, which they did in the French language. Several of these men owned slaves involved in the revolt.

152 They intended . . . planters' visions: In examining the court's motives, it is interesting to examine this court action through the lens

of scholarship about later revolts. The court involved in suppressing the Denmark Vesey conspiracy in Charleston in 1822 had as its first priority to stop the insurrection, with justice being a lesser aim. "It acted on the premise that it must suppress an impending slave insurrection, and it interrogated witnesses, passed judgment, and pronounced sentences accordingly," wrote historian Michael Johnson. M. P. Johnson, "Denmark Vesey and His Co-Conspirators," 942.

152 "to judge the rebel slaves . . . promptly destroyed": "Summary of Trial Proceedings of Those Accused of Participating in the Slave Uprising of January 9, 1811," trans. and ed. Dormon, "Notes and Documents."

153 "perfectly knew": Andry to Claiborne, January 11, 1811, in Thrasher, *On to New Orleans!*, 268.

153 "The confessions . . . an infuriated crowd": *North Carolina, Supreme Court, Raleigh: State v. George (a slave)*, June 1858 Manuscript Case File No. 7559, in Jones, *Reports of Cases at Law* 50:233–36.

154 "principal chief of the brigands . . . that of Mr. Reine, the older": Denunciations.

155 "he admitted . . . denounce anyone": Denunciations.

155 Some slaves . . . a different opinion: Denunciations; American Uprising Slave Database.

156 "confessed his guilt . . . ability to speak": Denunciations.

156 "These rebels . . . etc., etc., etc.": Conrad, *The German Coast*, 102.

156 eleven separate leaders: Amar of the Charbonnet plantation; Baptiste of the Bernoudy plantation; Jean of the Arnauld plantation; Harry of the Kenner and Henderson plantation; Zenon, Pierre, and Dagobert of the Delhomme plantation; Eugene of the Labranche plantation; Kook and Quamana of the James Brown plantation; and Charles Deslondes of the Deslondes plantation were all leaders of the uprising. Denunciations.

156 These leaders . . . from white fathers: Harry and Charles were mulattos. Eugene was a Louisiana Creole. Pierre was Kongolese. Kook and Quamana had only recently arrived in Louisiana from Africa. Hall, Louisiana Slave Database.

156 Their names . . . Anglo-American: Harry had an Anglo-American

name and came from a plantation owned by Americans. Charles, Jean, Pierre, and Baptiste had French names, and belonged to French planters. Quamana and Kook are anglicizations of West African names. Zenon was a Spanish name, while Dagobert was German. Hall, Louisiana Slave Database.

157 "GREAT EXAMPLE . . . public tranquility" . . . "in accordance with the authority . . . tranquility in the future": Conrad, *The German Coast*, 102.

158 The first floor . . . brought for trial: Bernhard, *Travels Through North America*, 2:60.

158. "It is presumed . . . be acquitted": John Shaw to Paul Hamilton, New Orleans, 18 January 1811.

158 Though Jean was found guilty . . . public official: *New Orleans City Court, Case No. 187*, January 17, 1811, in Thrasher, *On to New Orleans!*, 235.

158 The court treated . . . mercy from the court: *New Orleans City Court, Case No. 192*, January 21 1811, in Thrasher, *On to New Orleans!*, 242.

158 The court commuted . . . recent insurrection: *New Orleans City Court, Case No. 192*, January 18, 1811, in Albert Thrasher, *On to New Orleans!*, 241.

158 swayed on the levees in front of their masters' plantations: *New Orleans City Court, Case No. 184*, January 16, 1811, in Thrasher, *On to New Orleans!*, 231.

158 "hung at the usual place in the City of New Orleans": *New Orleans City Court, Case No. 185*, January 16, 1811, in Thrasher, *On to New Orleans!*, 233.

158 lower gates of the city: *New Orleans City Court, Case No. 188*, January 17, 1811, in Thrasher, *On to New Orleans!*, 237.

159 "It is just . . . of the guilty": Claiborne to John N. Destrehan, New Orleans, January 16, 1811, in W. C. C. Claiborne, *Official Letter Books*, 100–101.

160 "mischief" . . . "only": Claiborne to Doctor Steele, New Orleans, January 20, 1811, in W. C. C. Claiborne, *Official Letter Books*, 112–13.

160 "There can be . . . within it": Sprague, *The North Eastern Boundary Controversy*, 89–90.

160 The court system . . . body politic: The courts were the most immediate manifestation of that power, the most tangible embodiment of American government. "As agents of Americanization, county

judges and justices of the peace presided over the day-to-day application of American judicial practices on the most basic levels of the legal system, the local courts," wrote legal historian Mark Fernandez. "These inferior courts represented in the territory, as elsewhere in the republic, the one agency of the government that most likely touched ordinary citizens in the routine course of their daily lives." Mark Fernandez, "Local Justice in the Territory of Orleans, W.C.C. Claiborne's Courts: Judges and Justices of the Peace," in Fernandez and Billings, *A Law Unto Itself?*, 97.

161 "We are sorry . . . let them govern": Thompson, "National Newspaper and Legislative Reactions," 15.

161 "no doubt exists" . . . United States Army: *National Intelligencer*, February 19, 1811.

13. FRIENDS OF NECESSITY

167 He informed them . . . to the German Coast: W. C. C. Claiborne, *Official Letter Books*, 100–104.

168 "It seemed . . . conviction of others": N. H. Claiborne, *Notes on the War*, 105.

168 "warmest congratulations": W. C. C. Claiborne, *Official Letter Books*, 121.

169 Though he did not mention it . . . and Canada: Hatfield, *William Claiborne*, 323.

169 "The late daring . . . neighboring plantations": W. C. C. Claiborne, *Official Letter Books*, 123.

169 "made an impression . . . be effaced": W. C. C. Claiborne, *Official Letter Books*, 123.

171 "The faithful Citizens . . . well-regulated Militia": W. C. C. Claiborne, *Official Letter Books*, 124.

171 "our Security . . . the Militia": Magloire Guichard, "Answer, of the House of Representatives to Governor Claiborne's Speech," in W. C. C. Claiborne, *Official Letter Books*, 130.

171 "late unfortunate Insurrection . . . of the Militia": Jean Noël Destrehan, "Answer, of the Legislative Council to Governor Claiborne's Speech" in W. C. C. Claiborne, *Official Letter Books*, 127.

171 "awful lesson . . . lately quelled": Junius Rodriguez, "Always 'En

Garde': The Effects of Slave Insurrection upon the Louisiana Mentality, 1811–1815," in Labbé, *Louisiana: The Purchase and Its Aftermath*, 402–3.

172 The militia . . . train and organize: Rodriguez, "Always 'En Garde'," in Labbé, *Louisiana: The Purchase and Its Aftermath*, 402.

172 "the state of the population . . . other weighty considerations": "A Message from the Legislative Council to Pres. James Madison," Louisiana *Courier*, February 8, 1811, in Thrasher, *On to New Orleans!*, 271.

173 "Feeling that our destiny . . . Ark of safety": W. C. C. Claiborne, *Official Letter Books*, 131.

173 three additional gunboats: Rodriguez, "Always 'En Garde'," in Labbé, *Louisiana: The Purchase and Its Aftermath*, 403.

174 the mayor sent a message . . . "are responsible": Official Proceedings, New Orleans City Council, in Thrasher, *On to New Orleans!*, 275.

174 "Act . . . Territory": *L'Ami des Lois*, New Orleans, February 7, 1811.

175 "[The average resident] will not embody . . . unites society": *Louisiana Gazette* (New Orleans), April 1, 1811.

175 planters filed claims for about a third of the slaves: Conrad, *The German Coast*, 107–10.

175 "It is a fact of notoriety . . . easily anticipated": W. Claiborne, "Speech. Delivered by Governor Claiborne to both Houses of the Legislative Body of the Territory of Orleans," January 29, 1811, in W. C. C. Claiborne, *Official Letter Books*, 123.

14. STATEHOOD AND THE YOUNG AMERICAN NATION

177 "Strange as it may seem . . . almost commingled": Northup/Eakin and Logsdon, *Twelve Years a Slave*, 23.

177 "discouraging foreign intrigues" . . . "internal discontent": E. S. Brown, *Constitutional History of the Louisiana Purchase*, 190.

178 "The public . . . *white fellow citizens*": Thompson, "National Newspaper and Legislative Reactions," 17.

179 The combined population . . . in the United States: Rothman, *Slave Country*, 221.

179 The slave population . . . American Revolution and 1820: Rothman, *Slave Country*, ix–x.

180 "perfidious Britons . . . in its defense": Hatfield, *William Claiborne*, 290.

181 "the officer Commanding the English Fleet . . . with black troops": W. C. C. Claiborne, *Official Letter Books*, volume 6, 282.

181 "a powerful savage and negro army . . . devoted country": Rodriguez, "Always 'En Garde'," in Labbé, *Louisiana: The Purchase and Its Aftermath*, 410.

181 "like blades of grass . . . before the whirlwind": Hatfield, *William Claiborne*, 297.

182 a British fort at Prospect Bluff: Owsley and Smith, *Filibusters and Expansionists*, 104–5.

182 300 black men, women, and children: Meltzer, *Hunted Like a Wolf*, 50.

182 "I have little doubt . . . on which it stands": Giddings, *The Exiles of Florida*, 36–37.

183 In July of 1815 . . . "stolen negroes": Giddings, *The Exiles of Florida*, 36–37, 42–43.

183 The Spanish controlled . . . invaded Pensacola: Owsley and Smith, *Filibusters and Expansionists*, 159–60.

183 "an imaginary line in the woods": Israel, *State of the Union Messages*, I: 156–65.

185 "The invasion . . . American soil": Richardson, *A Compilation of the Messages and Papers*, 6:2290–92.

15. THE SLAVES WIN THEIR FREEDOM

188 By February . . . steam gunboats: Winters, *The Civil War in Louisiana*, 85.

188 Two well-armed . . . Louisiana vessels: Winters, *The Civil War in Louisiana*, 89.

189 But Farragut . . . in the night: Winters, *The Civil War in Louisiana*, 85–86.

189 On April 18 . . . Fort Jackson's guns: Winters, *The Civil War in Louisiana*, 88.

190 "People were amazed . . . speechless astonishment": Winters, *The Civil War in Louisiana*, 96.

190 "To the negroes . . . hour of triumph": Roland, *Louisiana Sugar Plantations During the Civil War*, 48–49.

190 "like thrusting a walking stick into an ant-hill": Roland, *Louisiana Sugar Plantations During the Civil War*, 92–93.

191 "Revolt & Insurrection . . . Lincoln and Freedom": Rodrigue, *Reconstruction in the Cane Fields*, 36.

191 A planter just outside . . . the plantation economy: Ripley, *Slaves and Freedmen in Civil War Louisiana*, 17.

191 In August of 1862 . . . were captured: Roland, *Louisiana Sugar Plantations During the Civil War*, 97.

191 "I shall treat . . . I fancy": Ripley, *Slaves and Freedmen in Civil War Louisiana*, 33.

192 "Any attempt . . . in the end": McPherson, *The Negro's Civil War*, 17–18.

193 "in time of actual armed rebellion . . .City of New Orleans": Richardson, *A Compilation of the Messages and Papers*, 5:3359.

193 At Mooreland plantation . . . "Glory to Abe Lincoln": Roland, *Louisiana Sugar Plantations During the Civil War*, 98–100.

193 "the slave population . . . of our families": Ripley, *Slaves and Freedmen in Civil War Louisiana*, 97.

194 "If we hadn't . . . naturally manhood": McPherson, *The Negro's Civil War*, 213.

194 "One morning the bell . . . away to the woods": Berlin et al., *Free at Last*, 51–52.

195 the first black regiments . . . by Union troops: Redkey, *A Grand Army of Black Men*, 3–5.

195 Initially . . . cowardice: McPherson, *The Negro's Civil War*, 163–64.

195 "There is not one man . . . field of battle": McPherson, *The Negro's Civil War*, 163.

195 Over the next two years . . . major battles: McPherson, *The Negro's Civil War*, 237.

196 "We can not spare . . . Union goes with it": McPherson, *The Negro's Civil War*, 235.

196 "We are now making up . . . root neor branch": Berlin et al., *Free at Last*, 482.

197 "visited plantations" . . . "called their wives": Berlin et al., *Free at Last*, 112.

197 "the Federal Government . . . the existing states": McPherson, *The Negro's Civil War*, 3–4.

198 "were so far inferior . . . reduced to slavery": Stauffer, *Giants*, 157.

16. THE COVER-UP

200 "It has been documented . . . provided for": Brochure on Destrehan Plantation Web site.

201 "Hidden History Tours . . . But not anymore!": Leon Waters, "Tours."

202 By 1996 . . . revolutionary struggle: Thrasher, *On to New Orleans!*, 48.

203 "overthrow their oppressors . . . 'white' rulers": Thrasher, *On to New Orleans!*, 48.

203 "On to New Orleans!" . . . "Freedom or death!": Thrasher, *On to New Orleans!*, 51.

203 "patently in contradiction with the truth": Thrasher, *On to New Orleans*, 65.

203 "This revolt stimulated . . . never abate": Thrasher, *On to New Orleans!*, 66.

203 "sham . . . capitalist moneybags": Thrasher, *On to New Orleans!*, 1.

205 "This incident . . . subjugated": Gayarré, *History of Louisiana*, 4:267.

206 "high circular gallery . . . kept them at bay": Gayarré, *History of Louisiana*, 4:267–68.

206 In 1918 . . . to the uprising: U. B. Phillips, *American Negro Slavery*, 474.

206 "were largely deprived . . . so strongly gives": U. B. Phillips, *American Negro Slavery*, 454.

207 "It is a land . . . a white man's country": U. B. Phillips, "The Central Theme of Southern History."

207 "growling" . . . "They must know who were their masters": Kendall, "Shadow over the City," 144–47.

207 through the work of Communist activists: Gilmore, *Defying Dixie*, 6.

208 "saw [Communism] . . . race relations": *New York Times*, March 20, 2003.

208 With Hitler gaining power . . . dangerous ideologies: Gilmore, *Defying Dixie*, 197–99.

209 "Ulrich B. Phillips . . .'by racial quality submissive'": Aptheker, *American Negro Slave Revolts*, 13.

209 Aptheker devoted a short paragraph to the 1811 uprising: Aptheker, *American Negro Slave Revolts*, 98.

EPILOGUE

211 "You really think. . . half-naked?": Tyson, *Radio Free Dixie*, 84.

212 "a nigger . . . a funeral": Tyson, *Radio Free Dixie*, 86.

212 "We got our own M-1's . . . we had everything": Tyson, *Radio Free Dixie*, 88.

213 "When we started firing . . . come back": Tyson, *Radio Free Dixie*, 88–89.

214 "We get no justice . . .resort to that method": Tyson, *Radio Free Dixie*, 149.

215 "No sincere leader . . . by example": Tyson, *Radio Free Dixie*, 246.

216 "Now that the United States . . . for this undertaking": Tyson, *Radio Free Dixie*, 241.

216 white Southerners lynched close to 4,000 African Americans: University of Missouri–Kansas City School of Law faculty project Web site.

BIBLIOGRAPHY

Primary Sources

Acts Passed at the Second Session of the Third Legislature of the Territory of Orleans: Begun and Held in the City of New-Orleans, on Monday, the Twenty-Third Day of January in the Year of Our Lord One Thousand Eight Hundred and Eleven. New Orleans.: Thierry, 1811.

Aime, Valcour. *Plantation Diary of the Late Mr. Valcour Aime, Formerly Proprietor of the Plantation Known as the St. James Sugar Refinery, Situated in the Parish of St. James, and Now Owned by Mr. John Burnside.* New Orleans: Clark & Hofeline, 1878.

Bernhard, Duke of Saxe-Weimar-Eisenach. *Travels Through North America, During the Years 1825 and 1826.* 2 volumes. Philadelphia: Carey, Lea & Carey, 1828.

Brackenridge, H. M. *Views of Louisiana. Together with a Journal of a Voyage Up the Missouri River, in 1811.* Chicago: Quadrangle Books, 1962.

Claiborne, Nathaniel Herbert. *Notes on the War in the South; with Biographical Sketches of the Lives of Montgomery, Jackson, Sevier, The Late Gov. Claiborne, and Others.* Richmond, Va.: William Ramsay, 1819.

Claiborne, William C. C. *Official Letter Books of W. C. C. Claiborne, 1801–1816.* Edited by Dunbar Rowland. Jackson, Miss.: State Department of Archives and History, 1917.

Conrad, Glenn R. *The German Coast: Abstracts of the Civil Records of St. Charles and St. John the Baptist Parishes, 1804–1812.* Lafayette: Center for Louisiana Studies, University of Southwestern Louisiana, 1981.

Dubois, Laurent, and John D. Garrigus. *Slave Revolution in the Caribbean, 1789–1804: A Brief History with Documents*. New York: Bedford/St. Martin's, 2006.

Engerman, Stanley, Seymour Drescher, and Robert Paquette. *Slavery*. New York: Oxford University Press, 2001.

Equiano, Olaudah. *The Interesting Narrative of the Life of Olaudah Equiano, or Gustavus Vassa, the African/Written by Himself; Authoritative Text, Contexts, Criticism*. Edited by Werner Sollors. New York: Norton, 2001.

Gayarré, Charles. *History of Louisiana*. 4 volumes. New York: Redfield, 1854–1866.

Jefferson, Thomas. *The Life and Writings of Thomas Jefferson*. Edited by S. E. Forman. New York: Braunworth, Munn & Barber, 1900.

——— *Notes on the State of Virginia: With Related Documents*. Edited by David Waldstreicher. New York: Palgrave, 2002.

Latrobe, Benjamin Henry. *Impressions Respecting New Orleans; Diary & Sketches, 1818–1820*. Edited by Samuel Wilson Jr. New York: Columbia University Press, 1951.

de Laussat, Pierre-Clément. *Memoirs of My Life to My Son During the Years 1803 and After, Which I Spent in Public Service in Louisiana as Commissioner of the French Government for the Retrocession to France of that Colony and for Its Transfer to the United States*. Baton Rouge: Published for the Historic New Orleans Collection by the Louisiana State University Press, 1978.

Martin, François-Xavier. *A General Digest of the Acts of the Legislatures of the Late Territory of Orleans and of the State of Louisiana and the Ordinances of the Governour Under the Territorial Government: Preceded by the Treaty of Cession, the Constitution of the United States, and of the State, with the Acts of Congress Relating to the Government of the Country and the Land Claims Therein*. New Orleans, La.: Printed by Peter K. Wagner, 1816.

Phillips, Wendell. *Speeches, Lectures, and Letters*. Boston: Lee and Shepard, 1894.

Redkey, Edwin, ed. *A Grand Army of Black Men: Letters from African-American Soldiers in the Union Army, 1861–1865*. New York: Cambridge University Press, 1992.

Report of Spanish Consul, January 13, 1811, Eusebio Bardari y Azara to Vicente Folch, February 6, 1811. Legajo 221a, Papeles de Cuba, Archivo General de Indias, Seville, Spain. Microfilm, Historic New Orleans Collection, New Orleans. Translated for the author by Gwendolyn Midlo Hall.

Richardson, James, ed. *A Compilation of the Messages and Papers of the Presidents*, 20 vols. and *Supplement*. New York: Bureau of National Literature, 1896–1917.

Shaw, John, to Paul Hamilton. New Orleans, 18 January 1811. Washington, D.C.: National Archives, Record Group 48, Microfilm 14.

St. Charles Parish, *Original Acts, Book 41*, 1811. Translated by Robert Paquette. Clinton, N.Y.: Hamilton College, 2008. Cited in notes as Denunciations.

Stedman, John Gabriel. *Stedman's Surinam: Life in Eighteenth-Century Slave Society*. Edited by Richard Price and Sally Price. Baltimore: Johns Hopkins University Press, 1992.

Stoddard, Amos. *Sketches, Historical and Descriptive of Louisiana*. Philadelphia: M. Carey, 1812 and New York: AMS Press, 1973.

Thrasher, Albert. *On to New Orleans! Louisiana's Heroic 1811 Slave Revolt*. 2nd ed. New Orleans, La: Cypress Press, 1996.

Secondary Sources

Articles

Bartle, Philip F. W. "Forty Days; The Akan Calendar." *Africa: Journal of the International African Institute* 48 (1): 80–84.

Brewer, W. M. Untitled review. *Journal of Negro History* 29, no. 1 (January 1944): 87–90.

Craven, Avery. Untitled review. *Journal of Economic History* 5, no. 1 (May 1945): 75–77.

Darnton, Robert. "An Early Information Society: News and the Media in Eighteenth-Century Paris." *American Historical Review* 105, no. 1 (2000): 1–35.

Dormon, James. "Notes and Documents." *Louisiana History* 17 (Fall 1977): 473.

———. "The Persistent Specter: Slave Rebellion in Territorial Louisiana." *Louisiana History* 18, no. 4 (1977): 389–404.

Edwards, Laura F. "Enslaved Women and the Law: Paradoxes of Subordination in the Post-Revolutionary Carolinas." *Slavery & Abolition* 26, no. 2 (2005): 305–323.

Hamilton, G. De Roulhac. Untitled review. *American Historical Review* 49, no. 3 (April, 1944): 504–506.

James, C.L.R., under the name J. Meyer. Untitled review. *Fourth International* 10, no. 11 (December 1949): 337–341. Accessed at http://www.marxists.org/archive/james-clr/works/1949/12/aptheker.htm.

Johnson, Michael P. "Denmark Vesey and His Co-Conspirators." *William and Mary Quarterly* 58, no. 4 (2001): 915–76.

Johnson, Walter. "On Agency." *Journal of Social History* 37, no. 1, Special Issue (2003): 113–124.

Kendall, John S. "Shadow over the City." *Louisiana Historical Quarterly* 22 (1939): 142–165.

Law, Robin. "'My Head Belongs to the King': On the Political and Ritual Significance of Decapitation in Pre-Colonial Dahomey." *Journal of African History* 30, no. 3 (1989): 399–415.

———. "Horses, Firearms, and Political Power in Pre-Colonial West Africa." *Past and Present* 72, no. 1 (1976): 112–32.

Leglaunec, Jeanne-Pierre. "Slave Migrations in Spanish and Early American Louisiana: New Sources and New Estimates." *Louisiana History*:46, no. 2 (2005): 185–209.

"Memoirs of Micah Taul," *Register of the Kentucky Historical Society* 27, no. 79 (January 1929), 356, quoted in William C. C. Claiborne, *Interim Appointment: W.C.C. Claiborne Letter Book, 1804–1805*.

O'Donovan, Susan. "Trunk Lines, Land Lines, and Local Exchanges: Operationalizing Slaves' Grapevine Telegraph." Faculty Working Paper, Harvard College, 2008.

Paquette, Robert L. "The Great Louisiana Slave Revolt of 1811 Reconsidered." *Historical Reflections* 35, no.1 (Spring 2009): 72–96.

Phillips, Ulrich B. "The Central Theme of Southern History." *American Historical Review* 34, no. 1 (1928): 30–43.

Rodriguez, Junius P. "Always 'En Garde': The Effects of Slave Insurrection upon the Louisiana Mentality, 1811–1815." *Louisiana History* 33, no. 4 (1992): 399–416.

Thompson, Thomas Marshall. "National Newspaper and Legislative Reactions to Louisiana's Deslondes Slave Revolt of 1811." *Louisiana History* 33, no. 1 (1992): 5–29.

Thornton, John K. "African Dimensions of the Stono Rebellion." *American Historical Review* 96, no. 4 (1991): 1101–13.

———. "African Soldiers in the Haitian Revolution." *Journal of Caribbean Studies* 25 (1991): 59–80.

Books

Accilien, Cécile, Jessica Adams, and Elmide Méléance, eds. *Revolutionary Freedoms: A History of Survival, Strength and Imagination in Haiti.* Coconut Creek, Fla.: Caribbean Studies Press, 2006.

Akers, Donna. *Living in the Land of Death: The Choctaw Nation, 1830–1860.* East Lansing: Michigan State University Press, 2004.

Anderson, Benedict R. O'G. *Imagined Communities: Reflections on the Origin and Spread of Nationalism.* New York : Verso, 2006.

Aptheker, Herbert. *American Negro Slave Revolts.* New York: Columbia University Press, 1943.

———. *Nat Turner's Slave Rebellion: Including the 1831 "Confessions".* Mineola, N.Y.: Dover Publications, 2006.

Armitage, David, and Michael J. Braddick, eds. *The British Atlantic World, 1500–1800.* New York: Palgrave Macmillan, 2002.

Arthur, Stanley Clisby. *The Story of the West Florida Rebellion.* St. Francisville, La.: The St. Francisville Democrat, 1935.

Baker, James Thomas. *Nat Turner: Cry Freedom in America.* Fort Worth: Harcourt Brace, 1998.

Bakhtin, M. M. *Voprosy Literatury i Ėstetiki (The Dialogic Imagination: Four Essays).* Translated by Caryl Emerson and Michael Holquist. Edited by Michael Holquist. Austin: University of Texas Press, 2004.

Barry, John M. *Rising Tide: The Great Mississippi Flood of 1927 and How It Changed America.* New York: Simon & Schuster, 1997.

Bender, Thomas, ed. *Rethinking American History in a Global Age.* Berkeley: University of California Press, 2002.

Berlin, Ira. *Many Thousands Gone: The First Two Centuries of Slavery in North America.* Cambridge, Mass.: Belknap Press of Harvard University Press, 1998.

Berlin, Ira, et al., eds. *Free at Last: A Documentary History of Slavery, Freedom, and the Civil War.* New York: The New Press, 1992.

Brown, Everett Somerville, *The Constitutional History of the Louisiana Purchase, 1803–1812.* Berkeley: University of California Press, 1920.

Brown, Vincent. *The Reaper's Garden: Death and Power in the World of Atlantic Slavery.* Cambridge, Mass.: Harvard University Press, 2008.

Buchanan, Thomas C. *Black Life on the Mississippi: Slaves, Free Blacks, and the Western Steamboat World.* Chapel Hill: University of North Carolina Press, 2004.

Carretta, Vincent. *Equiano, the African: Biography of a Self-Made Man.* Athens: University of Georgia Press, 2005.

Césaire, Aimé. *Lyric and Dramatic Poetry, 1946–82.* Translated by Clayton Eshleman and Annette Smith. Charlottesville: University Press of Virginia, 1990.

Childs, Matt D. *The 1812 Aponte Rebellion in Cuba and the Struggle Against Atlantic Slavery.* Chapel Hill: University of North Carolina Press, 2006.

Claiborne, William C. C. *Interim Appointment: W. C. C. Claiborne Letter Book, 1804–1805.* Edited by Jared William Bradley. Baton Rouge: Louisiana State University Press, 2002.

Conrad, Joseph. *Heart of Darkness and The Secret Sharer.* New York: Signet Classic 2008.

da Costa, Emília Viotti. *Crowns of Glory, Tears of Blood: The Demerara Slave Rebellion of 1823.* New York: Oxford University Press, 1994.

Crété, Liliane. *Daily Life in Louisiana, 1815–1830.* Translated by Patrick Gregory. Baton Rouge: Louisiana State University Press, 1981.

Davis, David Brion. *Inhuman Bondage: The Rise and Fall of Slavery in the New World.* New York: Oxford University Press, 2006.

Deerr, Noël. *The History of Sugar.* London: Chapman and Hall, 1949–50.

Du Bois, W. E. B. *Black Reconstruction in America.* Edited by David L. Lewis. New York: Simon & Schuster, 1995.

Dubois, Laurent. *Avengers of the New World: The Story of the Haitian Revolution.* Cambridge, Mass.: Belknap Press of Harvard University Press, 2004.

Egerton, Douglas R. *He Shall Go Out Free: The Lives of Denmark Vesey.* Madison, Wisc.: Madison House, 1999.

———. *Gabriel's Rebellion: The Virginia Slave Conspiracies of 1800 and 1802.* Chapel Hill: University of North Carolina Press, 1993.

Fernandez, Mark F., and Warren M. Billings, eds. *A Law unto Itself? Essays in the New Louisiana Legal History.* Baton Rouge: Louisiana State University Press, 2001.

Fields, Barbara Jeanne. *Slavery and Freedom on the Middle Ground: Maryland During the Nineteenth Century*. New Haven, Conn.: Yale University Press, 1985.

Follett, Richard J. *The Sugar Masters: Planters and Slaves in Louisiana's Cane World, 1820–1860*. Baton Rouge: Louisiana State University Press, 2005.

Foner, Eric. *Nat Turner*. Englewood Cliffs, N.J.: Prentice-Hall, 1971.

Foucault, Michel. *Discipline and Punish: The Birth of the Prison*. 2nd edition. Translated by Alan Sheridan. New York: Vintage Books, 1995.

Fox-Genovese, Elizabeth, and Eugene D. Genovese. *Fruits of Merchant Capital: Slavery and Bourgeois Property in the Rise and Expansion of Capitalism*. New York: Oxford University Press, 1983.

Freehling, William W. *The Road to Disunion*. New York: Oxford University Press, 1990–2007.

Fynn, John Kofi. *Asante and Its Neighbours, 1700–1807*. Evanston, Ill.: Northwestern University Press, 1971.

Gaspar, David Barry, and David Patrick Geggus, eds. *A Turbulent Time: The French Revolution and the Greater Caribbean*. Bloomington: Indiana University Press, 1997.

Geer, Curtis Manning. *The Louisiana Purchase and the Westward Movement*. The History of North America, 8. Philadelphia: George Barrie & Sons, 1904.

Geggus, David Patrick, ed. *The Impact of the Haitian Revolution in the Atlantic World*. Columbia: University of South Carolina Press, 2001.

Genovese, Eugene D. *From Rebellion to Revolution: Afro-American Slave Revolts in the Making of the Modern World*. Baton Rouge: Louisiana State University Press, 1979.

———. *Roll, Jordan, Roll: The World the Slaves Made*. New York: Vintage Books, 1976.

Giddings, Joshua R. *The Exiles of Florida; Or, The Crimes Committed by Our Government Against the Maroons Who Fled from South Carolina and Other Slave States, Seeking Protection Under Spanish Laws*. A facsimile reproduction of the 1858 edition. Gainesville: University of Florida Press, 1964.

Gilmore, Glenda Elizabeth. *Defying Dixie: The Radical Roots of Civil Rights, 1919–1950*. New York: W. W. Norton & Co., 2008.

Greeley, Horace. *The American Conflict: A History of the Great Rebellion in the United States of America, 1860-'64*. Hartford, Conn.: O. D. Case & Company, 1865.

Hahn, Steven. *A Nation Under Our Feet: Black Political Struggles in the Rural South from Slavery to the Great Migration.* Cambridge, Mass.: Belknap Press of Harvard University Press, 2003.

Hall, Gwendolyn Midlo. *Africans in Colonial Louisiana: The Development of Afro-Creole Culture in the Eighteenth Century.* Baton Rouge: Louisiana State University Press, 1992.

Harms, Robert W. *The Diligent: A Voyage Through the Worlds of the Slave Trade.* New York: Basic Books, 2002.

Harvey, Horace H., Katherine Harvey Roger, and Louise Destrehan Roger D'Oliveira, *To Reach Afar: Memoirs and Biography of the Destrehan and Harvey Families of Louisiana.* Clearwater, Fla: Hercules Pub. Co., 1974.

Hatfield, Joseph T. *William Claiborne: Jeffersonian Centurion in the American Southwest.* Lafayette: University of Southwestern Louisiana, 1976.

Haywood, Ian. *Bloody Romanticism: Spectacular Violence and the Politics of Representation, 1776–1832.* New York: Palgrave Macmillan, 2006.

Heidler, David Stephen, and Jeanne T. Heidler. *The War of 1812.* Westport, Conn.: Greenwood Press, 2002.

Hess, Earl. *The Union Soldier in Battle: Enduring the Ordeal of Combat.* Lawrence: University Press of Kansas, 1997.

Heywood, Linda Marinda, ed. *Central Africans and Cultural Transformations in the American Diaspora.* New York: Cambridge University Press, 2002.

Hickey, Donald R. *The War of 1812: A Forgotten Conflict.* Urbana: University of Illinois Press, 1989.

Higginson, Thomas Wentworth. *Nat Turner.* Los Angeles: Vanguard Society of America, 1962.

Hunt, Alfred N. *Haiti's Influence on Antebellum America: Slumbering Volcano in the Caribbean.* Baton Rouge: Louisiana State University Press, 1988.

Ingersoll, Thomas N. *Mammon and Manon in Early New Orleans: The First Slave Society in the Deep South, 1718–1819.* Knoxville: University of Tennessee Press, 1999.

Israel, Fred L., ed., *The State of the Union Messages of the Presidents, 1790–1966.* 3 volumes. New York: Chelsea House, 1966.

James, C.L.R. *The Black Jacobins: Toussaint L'Ouverture and the San Domingo Revolution.* 2nd ed. New York: Vintage Books, 1989.

Johnson, F. Roy. *The Nat Turner Story: History of the South's Most Important Slave Revolt, with New Material Provided by Black Tradition and White Tradition.* Murfreesboro, N.C.: Johnson Pub. Co., 1970.

Johnson, Walter. *Soul by Soul: Life Inside the Antebellum Slave Market.* Cambridge, Mass.: Harvard University Press, 1999.

Jones, Hamilton C. *Reports of Cases at Law Argued and Determined in the Supreme Court of North Carolina.* Volume 5. Raleigh, N.C.: S. Gales, 1854–60.

Jordan, Winthrop D. *Tumult and Silence at Second Creek: An Inquiry into a Civil War Slave Conspiracy.* Revised edition. Baton Rouge: Louisiana State University Press, 1995.

Joyce, Patrick. *The Rule of Freedom: Liberalism and the Modern City.* New York: Verso, 2003.

Kane, Harnett Thomas. *Gone Are the Days: An Illustrated History of the Old South.* New York: Dutton, 1960.

———. *Natchez on the Mississippi.* New York: W. Morrow and Company, 1947.

———. *Plantation Parade: The Grand Manner in Louisiana.* New York: W. Morrow and Company, 1945.

Kastor, Peter J. *The Nation's Crucible: The Louisiana Purchase and the Creation of America.* New Haven: Yale University Press, 2004.

Kaye, Anthony E. *Joining Places: Slave Neighborhoods in the Old South.* Chapel Hill: University of North Carolina Press, 2007.

Kennedy, Roger G. *Mr. Jefferson's Lost Cause: Land, Farmers, Slavery, and the Louisiana Purchase.* New York: Oxford University Press, 2003.

Kinser, Sam. *Carnival, American Style: Mardi Gras at New Orleans and Mobile.* Chicago: University of Chicago Press, 1990.

Kiple, Kenneth F. *The Caribbean Slave: A Biological History.* New York: Cambridge University Press, 1984.

Labbé, Delores, ed. *Louisiana: The Purchase and Its Aftermath, 1800–1830.* Louisiana Purchase Bicentennial Series in Louisiana History, volume 3. Lafayette.: Center for Louisiana Studies, 1998.

Langley, Lester D. *The Americas in the Age of Revolution, 1750–1850.* New Haven, Conn.: Yale University Press, 1996.

Latour, Bruno. *Reassembling the Social: An Introduction to Actor-Network-Theory.* New York: Oxford University Press, 2005.

Lepore, Jill. *The Name of War: King Philip's War and the Origins of American Identity.* New York: Knopf, 1998.

———. *New York Burning: Liberty, Slavery, and Conspiracy in Eighteenth-Century Manhattan.* New York: Knopf, 2005.

Lewis, Berkeley R. *Small Arms and Ammunition in the United States Service.* Washington, D.C.: Smithsonian Institution, 1956.

Libby, David J. *Slavery and Frontier Mississippi, 1720–1835.* Jackson: University Press of Mississippi, 2004.

Linebaugh, Peter, and Marcus Buford Rediker. *The Many-Headed Hydra: Sailors, Slaves, Commoners, and the Hidden History of the Revolutionary Atlantic.* Boston: Beacon Press, 2000.

Loewen, James W. *Lies My Teacher Told Me: Everything Your American History Textbook Got Wrong.* New York: Simon & Schuster, 2007.

Lowell, James Russell. *The Poetical Works of James Russell Lowell.* Boston: Houghton, Mifflin, 1890.

McCoy, Drew R. *The Elusive Republic: Political Economy in Jeffersonian America.* Chapel Hill: Published for the Institute of Early American History and Culture, Williamsburg, Va., by the University of North Carolina Press, 1980.

McDonald, Roderick A. *The Economy and Material Culture of Slaves: Goods and Chattels on the Sugar Plantations of Jamaica and Louisiana.* Baton Rouge: Louisiana State University Press, 1993.

McKay, Claude. *Harlem Shadows: The Poems of Claude McKay.* New York: Harcourt Brace, 1922.

McKivigan, John R., and Stanley Harrold, eds. *Antislavery Violence: Sectional, Racial, and Cultural Conflict in Antebellum America.* Knoxville: University of Tennessee Press, 1999.

McPherson, James M. *The Negro's Civil War: How American Blacks Felt and Acted During the War for the Union.* New York: Vintage Books, 2003.

Meinig, D. W. *The Shaping of America: A Geographical Perspective on 500 Years of History.* New Haven: Yale University Press, 1986.

Meltzer, Milton. *Hunted Like a Wolf: The Story of the Seminole War.* New York: Farrar, Straus and Giroux, 1972.

Miller, Joseph C. *Way of Death: Merchant Capitalism and the Angolan Slave Trade, 1730–1830.* Madison: University of Wisconsin Press, 1988.

Miller, Randall M., and John David Smith, eds. *Dictionary of Afro-American Slavery*. New York: Greenwood Press, 1988.

Mintz, Sidney. *Sweetness and Power: The Place of Sugar in Modern History*. New York: Penguin, 1985.

Moody, Vernie Alton. *Slavery on Louisiana Sugar Plantations*. New Orleans: Cabildo, 1924.

Mullin, Michael. *Africa in America: Slave Acculturation and Resistance in the American South and the British Caribbean, 1736–1831*. Urbana: University of Illinois Press, 1992.

Northup, Solomon. *Twelve Years a Slave*. Radford, Va.: Wilder Publications, 2008.

———. *Twelve Years a Slave*. Edited by Sue L. Eakin and Joseph Logsdon. Baton Rouge: Louisiana State University Press, 1968.

Oates, Stephen B. *The Fires of Jubilee: Nat Turner's Fierce Rebellion*. New York: HarperPerennial, 1990.

Owsley, Frank Lawrence, Jr., and Gene A. Smith. *Filibusters and Expansionists: Jeffersonian Manifest Destiny, 1800–1821*. Tuscaloosa: University of Alabama Press, 1997.

Palmié, Stephan, ed. *Slave Cultures and the Cultures of Slavery*. Knoxville: University of Tennessee Press, 1995.

Paquette, Robert L. *Sugar Is Made with Blood: The Conspiracy of La Escalera and the Conflict Between Empires Over Slavery in Cuba*. 1st edition. Middletown, Conn.: Wesleyan University Press, 1988.

Paquette, Robert L., and Louis A. Ferleger, eds. *Slavery, Secession, and Southern History*. Charlottesville: University Press of Virginia, 2000.

Patterson, Orlando. *Slavery and Social Death: A Comparative Study*. Cambridge, Mass.: Harvard University Press, 1982.

Phillips, Ulrich B. *American Negro Slavery: A Survey of the Supply, Employment and Control of Negro Labor as Determined by the Plantation Regime*. New York: D. Appleton and Company, 1918.

Rehder, John B. *Delta Sugar: Louisiana's Vanishing Plantation Landscape*. Baltimore: Johns Hopkins University Press, 1999.

Remini, Robert Vincent. *The Battle of New Orleans*. New York: Viking, 1999.

Ripley, C. Peter. *Slaves and Freedmen in Civil War Louisiana*. Baton Rouge: Louisiana State University Press, 1976.

Rodrigue, John C. *Reconstruction in the Cane Fields: From Slavery to Free Labor in Louisiana's Sugar Parishes, 1862–1880.* Baton Rouge: Louisiana State University Press, 2001.

Rodriguez, Junius P., ed. *Encyclopedia of Slave Resistance and Rebellion.* Westport, Conn.: Greenwood Press, 2007.

Roland, Charles Pierce. *Louisiana Sugar Plantations During the Civil War.* Leiden, Netherlands: E. J. Brill, 1957.

Rothman, Adam. *Slave Country: American Expansion and the Origins of the Deep South.* Cambridge, Mass.: Harvard University Press, 2005.

Scarry, Elaine. *The Body in Pain: The Making and Unmaking of the World.* New York: Oxford University Press, 1985.

Schwartz, Marie Jenkins. *Birthing a Slave: Motherhood and Medicine in the Antebellum South.* Cambridge, Mass.: Harvard University Press, 2006.

Scott, James C. *Domination and the Arts of Resistance: Hidden Transcripts.* New Haven: Yale University Press, 1990.

Seebold, Herman Boehm de Bachellé. *Old Louisiana Plantation Homes and Family Trees.* Gretna, La.: Pelican, 1941.

Sidbury, James. *Ploughshares into Swords: Race, Rebellion, and Identity in Gabriel's Virginia, 1730–1810.* New York: Cambridge University Press, 1997.

Sitterson, Joseph Carlyle. *Sugar Country: The Cane Sugar Industry in the South, 1753–1950.* Lexington: University of Kentucky Press, 1953.

Slotkin, Richard. *Gunfighter Nation: The Myth of the Frontier in Twentieth-Century America.* New York: Atheneum, 1992.

Smith, Philip Chadwick Foster, and G. Gouverneur Meredith S. Smith. *Cane, Cotton & Crevasses: Some Antebellum Louisiana and Mississippi Plantations of the Minor, Kenner, Hooke, and Shepherd Families.* Bath, Maine: Renfrew Group, 1992.

Spierenburg, Petrus Cornelis. *The Spectacle of Suffering: Executions and the Evolution of Repression: From a Preindustrial Metropolis to the European Experience.* New York: Cambridge University Press, 1984.

Sprague, John Francis. *The North Eastern Boundary Controversy and the Aroostook War.* Dover, Maine: The Observer Press, 1910.

Starobin, Robert S. *Denmark Vesey: The Slave Conspiracy of 1822.* Englewood Cliffs, N.J.: Prentice-Hall, 1970.

Stauffer, John. *Giants: The Parallel Lives of Frederick Douglass & Abraham Lincoln.* New York: Twelve, 2008.

Styron, William. *The Confessions of Nat Turner.* New York: Vintage Books, 1993.

Sublette, Ned. *The World That Made New Orleans: From Spanish Silver to Congo Square.* Chicago: Lawrence Hill Books: 2008.

Taylor, Joe Gray. *Negro Slavery in Louisiana.* New York: Negro Universities Press, 1969.

Thomas, David Y. *A History of Military Government in Newly Acquired Territory of the United States.* New York: Columbia University Press, 1904.

Trouillot, Michel-Rolph. *Silencing the Past: Power and the Production of History.* Boston: Beacon Press, 1995.

Tyson, Timothy B. *Radio Free Dixie: Robert F. Williams and the Roots of Black Power.* Chapel Hill: University of North Carolina Press, 1999.

Verdery, Katherine. *The Political Lives of Dead Bodies: Reburial and Postsocialist Change.* New York: Columbia University Press, 1999.

Weber, David J. *The Spanish Frontier in North America.* New Haven: Yale University Press, 1992.

Williams, Lorraine A. *Africa and the Afro-American Experience: Eight Essays.* Washington, D.C.: Howard University Press, 1977.

Winters, John David. *The Civil War in Louisiana.* Baton Rouge: Louisiana State University Press, 1963.

Yoes, Henry E. *Louisiana's German Coast: A History of St. Charles Parish.* Lake Charles, La.: Racing Pigeon Digest Pub. Co. 2005.

Young, Tommy Richard II. "The United States Army in the South, 1789–1835. (Volumes I and II)." PhD dissertation, Louisiana State University and Agricultural and Mechanical College, 1973.

Databases and Web Sites

American Uprising Slave Database. The database, created by the author, is a cross-reference of the St. Charles Parish Original Acts, encompassing the court trials and reimbursement claims translated by Glen Conrad, the trials from the City Court of New Orleans, as transcribed by Thrasher, and a set of runaway advertisements compiled by Thrasher.

Declaration of the Rights of Man. Accessed at The Avalon Project, Lillian Goldman Law Library, Yale Law School: http://avalon.law.yale.edu/18th_century/rightsof.asp.

Destrehan Plantation. www.destrehanplantation.org/pdf/destrehan brochure.pdf

Hall, Gwendolyn Midlo. Louisiana Slave Database 1719–1820, in Gwendolyn Midlo Hall, ed., *Afro-Louisiana History and Genealogy, 1699–1860*. CD-ROM. Baton Rouge: Louisiana University Press, 2000.

Summer Institute of Linguistics, Inc. (SIL) Aukan–English Dictionary. www.sil.org/americas/suriname/Aukan/English/AukanEngDict Index.html.

University of Missouri–Kansas City School of Law faculty project Web site, www.law.umkc.edu/faculty/projects/ftrials/shipp/lynchingyear .html.

Waters, Leon. Hidden History Tours. "Tours." www.historyhidden .com/tours (accessed February 12, 2009).

INDEX

Index

Private Games

BOOKS BY JAMES PATTERSON

The Private Novels

Private Games (with Mark Sullivan)
Private: #1 Suspect (with Maxine Paetro)
Private (with Maxine Paetro)

A complete list of books by James Patterson is at the back of this book. For previews and information about the author, visit JamesPatterson.com or find him on Facebook or at your app store.

Private Games

James Patterson
AND
Mark Sullivan

LITTLE, BROWN AND COMPANY
NEW YORK BOSTON LONDON

The characters and events in this book are fictitious. Any similarity to real persons, living or dead, is coincidental and not intended by the author.

Little, Brown and Company
Hachette Book Group
237 Park Avenue, New York, NY 10017
www.hachettebookgroup.com

First Edition: February 2012

Little, Brown and Company is a division of Hachette Book Group, Inc., and is celebrating its 175th anniversary in 2012. The Little, Brown name and logo are trademarks of Hachette Book Group, Inc.

Library of Congress Cataloging-in-Publication Data
Patterson, James
Private games : a novel / by James Patterson and Mark Sullivan.—1st ed.
p. cm.—(Private)
ISBN 978-0-316-20682-2 (hc) / 978-0-316-20680-8 (large print)
1. Olympic Games (30th : 2012 : London, England)—Fiction. 2. Private security services—Fiction. 3. Private investigators—Fiction. I. Sullivan, Mark T. II. Title.
PS3566.A822P765 2012
813'.54—dc23 2011035738

10 9 8 7 6 5 4 3 2 1

RRD-H

Printed in the United States of America

For Connor and Bridger, chasers of the Olympic dream

—M.S.

It is not possible with mortal mind to search out the purposes of the gods.

—Pindar

For then, in wrath, the Olympian thundered and lightninged, and confounded Greece.

—Aristophanes

Prologue

WEDNESDAY, JULY 25, 2012, 11:25 P.M.

THERE *ARE* SUPERMEN and superwomen who walk this earth.

I'm quite serious about that, and you can take me literally. Jesus Christ, for example, was a spiritual superman, as were Martin Luther and Gandhi. Julius Caesar was superhuman as well. So were Genghis Khan, Thomas Jefferson, Abraham Lincoln, and Adolf Hitler.

Think about scientists like Aristotle, Galileo, Albert Einstein, and J. Robert Oppenheimer. Consider artists like da Vinci, Michelangelo, and Vincent van Gogh, my favorite, who was so superior it drove him insane. Above all, don't forget athletically superior beings like Jim Thorpe, Babe Didrikson

Zaharias, and Jesse Owens; Larisa Latynina and Muhammad Ali; Mark Spitz and Jackie Joyner-Kersee.

Humbly, I include myself on this superhuman spectrum as well—and deservedly so, as you shall soon see.

In short, people like me are born for great things. We seek adversity. We seek to conquer. We seek to break through all limits—spiritually, politically, artistically, scientifically, and physically. We seek to right wrongs in the face of monumental odds. And we're willing to suffer for greatness, willing to engage in dogged effort and endless preparation with the fervor of a martyr—which, to my mind, is an exceptional trait in any human being at any age.

At the moment I have to admit that I'm certainly feeling exceptional, standing here in the garden of Sir Denton Marshall, a sniveling, corrupt old bastard if there ever was one.

Look at him on his knees, with his back to me and my knife at his throat.

Why, he trembles and shakes as if a stone had just clipped his head. Can you smell it? Fear? It surrounds him with an odor as rank as the air after a bomb explodes.

"Why?" he gasps.

"You've angered me, monster," I snarl at him, feeling a deeper-than-primal rage split my mind and seethe through every cell. "You've helped ruin the games, made them a mockery and an abomination."

"What?" he cries, acting bewildered. "What are you talking about?"

I deliver the evidence against him in three damning sentences that turn the skin of his neck livid and his carotid artery a sickening, pulsing purple.

"No!" he sputters. "That's...that's not true. You can't do this. Have you gone utterly mad?"

"Mad? Me?" I say. "Hardly. I'm the sanest person I know."

"Please," he says, tears rolling down his face. "Have mercy. I'm to be married on Christmas Eve."

My laugh is as caustic as battery acid. "In another life, Denton, I ate my own children. You'll get no mercy from me or my sisters."

As his confusion and horror become complete, I look up into the night sky, feeling storms rising in my head, and understanding once again that I *am* superior, superhuman, imbued with forces that go back thousands of years.

"For all true Olympians," I vow, "this act of sacrifice marks the beginning of the end of the modern games."

Then I wrench the old man's head back so his back arches.

And before he can scream, I furiously rip the blade across his throat with such force that his head comes free of his neck all the way to his spine.

Book One

THE FURIES

CHAPTER 1

THURSDAY, JULY 26, 2012, 9:24 A.M.

IT WAS MAD-DOG hot for London. Peter Knight's shirt and jacket were drenched with sweat as he sprinted north on Chesham Street past the Diplomat Hotel and skidded around the corner toward Lyall Mews in the heart of Belgravia, home to some of the most expensive real estate in the world.

Don't let it be true, Knight screamed internally as he entered the mews. *Dear God, don't let it be true.*

Then he saw a pack of newspaper reporters gathering at the yellow tape of a London Metropolitan Police barricade that blocked the road in front of a cream-colored Georgian-style townhome. Knight lurched to a stop, feeling like he was going to retch up the eggs and bacon he'd had for breakfast.

What would he ever tell Amanda?

Before Knight could compose his thoughts or still his stomach, his cell phone rang. He snatched it from his pocket without looking at caller ID.

"Knight," he managed to choke out. "That you, Jack?"

"No, Peter, it's Nancy," the voice replied in an Irish brogue. "Isabel has come down sick."

"What?" he groaned. "No...I just left the house an hour ago."

"She's running a temperature," the full-time nanny insisted. "I just took it."

"How high?"

"One hundred. She's complaining about her stomach, too."

"Lukey?"

"He seems fine," she said. "But—"

"Give them both a cool bath, and call me back if Isabel's temp hits a hundred and one," Knight said. He snapped the phone shut, swallowed the bile burning at the back of his throat.

A wiry man about six feet tall, with an appealing face and light brown hair, Knight had once been a special investigator assigned to the Old Bailey, home of England's Central Criminal Court. Two years ago, however, he joined the London office of Private International at twice the pay and prestige. Private has been called the Pinkerton Agency of the twenty-first century, with offices in every major city in the world staffed by top-notch forensic scientists, security specialists, and investigators such as Knight.

Compartmentalize, he told himself. *Be professional.* But this felt like the straw that would break the camel's back. Knight had already endured too much grief and loss, both personally and professionally. Just the week before, his boss, Dan Carter, and three of his colleagues had perished in a plane crash over the North Sea that was still under investigation. Could he live with another death?

Pushing that question and his daughter's illness to one side, Knight forced himself to hurry on through the sweltering heat toward the police barrier, giving the Fleet Street crowd a wide berth, and in so doing spotted Billy Casper, a Scotland Yard inspector he'd known for fifteen years.

He went straight to Casper, a blockish man with a pockmarked face who scowled the second he saw Knight. "Private's got no business in this, Peter."

"If that's Sir Denton Marshall dead in there, then Private does have business in this, and I do, too," Knight shot back forcefully. "Personal business, Billy. Is it Sir Denton?"

Casper said nothing.

"Is it?" Knight demanded.

Finally the inspector nodded, but he wasn't happy about it, and asked suspiciously, "How are you and Private involved?"

Knight stood there a moment, feeling lambasted by the news and wondering again how the hell he was going to tell Amanda. Then he shook off the despair and said, "The London Organising Committee for the Olympic Games is Private London's client. Which makes Sir Denton Private's client."

"And you?" Casper demanded. "What's your personal stake in this? You a friend of his or something?"

"Much more than a friend. He was engaged to my mother."

Casper's hard expression softened a bit and he chewed at his lip before saying, "I'll see if I can get you in. Elaine will want to talk to you."

Knight felt suddenly as if invisible forces were conspiring against him.

"Elaine caught this case?" he said, wanting to punch something. "You can't be serious."

"Dead serious, Peter," Casper said. "Lucky, lucky you."

CHAPTER 2

CHIEF INSPECTOR ELAINE Pottersfield was one of the finest detectives working for the Metropolitan Police, a twenty-year veteran of the force with a prickly, know-it-all style that got results. Pottersfield had solved more murders in the past two years than any other inspector at Scotland Yard. She was also the only person Knight knew who openly despised his presence.

An attractive woman in her forties, the inspector always put Knight in mind of a borzoi, with her large round eyes, aquiline face, and silver hair that cascaded about her shoulders. When he entered Sir Denton Marshall's kitchen, Pottersfield eyed him down her sharp nose, looking ready to bite at him if she got the chance.

"Peter," she said coldly.

"Elaine," Knight said.

"Not exactly my idea to let you into the crime scene."

"No, I imagine not," replied Knight, fighting to control his emotions, which were heating up by the second. Pottersfield always seemed to have that effect on him. "But here we are. What can you tell me?"

The Scotland Yard inspector did not reply for several moments. Then she finally said, "The maid found him an hour ago out in the garden, or what's left of him, anyway."

Flashing on memories of Sir Denton, the learned and funny man he'd come to know and admire over the past two years, Knight's legs felt wobbly, and he had to put his vinyl-gloved hand out on the counter to steady himself. "What's left of him?"

Pottersfield grimly gestured at the open French door.

Knight absolutely did not want to go out into the garden. He wanted to remember Sir Denton the last time he'd seen him, two weeks before, with his shock of startling white hair, scrubbed pink skin, and easy, infectious laugh.

"I understand if you'd rather not," Pottersfield said. "Inspector Casper said your mother was engaged to Sir Denton. When did that happen?"

"New Year's past," Knight said. He swallowed and moved toward the door, adding bitterly, "They were to be married on Christmas Eve. Another tragedy. Just what I need in my life, isn't it?"

Pottersfield's expression twisted in pain and anger, and she looked at the kitchen floor as Knight went by her and out into the garden.

Outside, the temperature was growing hotter. The air in the garden was still and stank of death and gore. On the flagstone terrace, five quarts of blood—the entire reservoir of Sir Denton's life—had run out and congealed around his decapitated corpse.

"The medical examiner thinks the job was done with a long curved blade that has a serrated edge," Pottersfield said.

Knight again fought off the urge to vomit. He tried to take the entire scene in, to burn it into his mind as if it were a series of photographs and not reality. Keeping everything at arm's length was the only way he knew to get through something like this.

Pottersfield said, "And if you look closely, you'll see some of the blood's been sprayed back toward the body with water from the garden hose. I'd expect the killer did it to wash away footprints and such."

Knight nodded, and then, by sheer force of will, moved his attention beyond the body, deeper into the garden, bypassing forensics techs gathering evidence from the flower beds and turning to a crime-scene photographer snapping away near the back wall.

Knight skirted the corpse by several feet and from that new perspective saw what the photographer was focusing on. It was from ancient Greece, and was one of Sir Denton's prized possessions: a headless limestone statue of an Athenian senator cradling a scroll and holding the hilt of a busted sword.

Sir Denton's head had been placed in the empty space between the statue's shoulders. His face was puffy, lax. His mouth was twisted to the left, as if he were spitting. And his eyes were open, dull, and, to Knight, shockingly forlorn.

For an instant, the Private operative wanted to break down. But then he felt himself swell with outrage. What kind of bar-

barian would do such a thing? And why? What possible reason could there be to behead Denton Marshall? The man was more than good. He was...

"You're not seeing it all, Peter," Pottersfield said behind him. "Go look at the grass in front of the statue."

Knight closed his hands to fists and walked off the terrace onto the grass, which scratched against the paper booties he wore over his shoes, making a sound that was as annoying to him as fingernails on a chalkboard. Then he saw it and stopped cold.

Five interlocking rings, the symbol of the Olympic Games, had been spray-painted on the grass.

Through the symbol, an *X* had been smeared in blood.